Lecture Notes in Computer Science   1089

Edited by G. Goos, J. Hartmanis and J. van Leeuwen

Advisory Board: W. Brauer   D. Gries   J. Stoer

W0051420

Springer

Berlin
Heidelberg
New York
Barcelona
Budapest
Hong Kong
London
Milan
Paris
Santa Clara
Singapore
Tokyo

G. Ramalingam

Bounded
Incremental Computation

 Springer

Series Editors

Gerhard Goos, Karlsruhe University, Germany

Juris Hartmanis, Cornell University, NY, USA

Jan van Leeuwen, Utrecht University, The Netherlands

Author

G. Ramalingam
IBM T.J. Watson Research Center
P.O. Box 704, Yorktown Heights, NY 10598, USA

Cataloging-in-Publication data applied for

Die Deutsche Bibliothek - CIP-Einheitsaufnahme

Ramalingam, G.:
Bounded incremental computation / G. Ramalingam. - Berlin ;
Heidelberg ; New York ; Barcelona ; Budapest ; Hong Kong ;
London ; Milan ; Paris ; Santa Clara ; Singapore ; Tokyo :
Springer, 1996
  (Lecture notes in computer science ; 1089)
  ISBN 3-540-61320-X
NE: GT

CR Subject Classification (1991): F.2, G.2, F.1, D.4.7, D.2.6, D.3.4

ISBN 3-540-61320-X Springer-Verlag Berlin Heidelberg New York

This work is subject to copyright. All rights are reserved, whether the whole or part of the material is
concerned, specifically the rights of translation, reprinting, re-use of illustrations, recitation, broadcasting,
reproduction on microfilms or in any other way, and storage in data banks. Duplication of this publication
or parts thereof is permitted only under the provisions of the German Copyright Law of September 9, 1965,
in its current version, and permission for use must always be obtained from Springer - Verlag. Violations are
liable for prosecution under the German Copyright Law.

© Springer-Verlag Berlin Heidelberg 1996
Printed in Germany

Typesetting: Camera-ready by author
SPIN 10513144   06/3142 – 5 4 3 2 1 0   Printed on acid-free paper

to

my mother, my father, and Raji

with love

என் தாய், என் தந்தை, ராஜி மூவருக்கும்

அன்புடன்

Preface

Algorithms and programs transform their input into their output. *Incremental computation* concerns the re-computation of output after a change in the input. An incremental algorithm, consequently, transforms a *change in input* into a *change in output*. Incremental algorithms, also called dynamic or on-line algorithms, are becoming increasingly important given the popularity of interactive systems, which must respond efficiently to a user's actions, which are usually "modifications" of an "input document".

In the context of incremental computation, small changes in the input are likely to cause correspondingly small changes in the output. It is natural to attempt to identify the part of the previous output that is no longer "correct" and "update" it. Where it is not possible to identify the affected part of the output *exactly*, an incremental algorithm may attempt to identify a *conservative approximation* (that is, an over-estimation) of the affected part of the output. Given some part of the output that needs to be recomputed, an incremental algorithm would benefit by processing only the portion of the input it needs to process in order to generate that part of the output. The effectiveness or efficiency of an incremental algorithm depends on how accurate an approximation to the affected region it can identify and on the overhead it incurs in doing this.

This book presents results — upper bound results, lower bound results, and experimental results — for several incremental computation problems. What is common to all these results is that we seek to determine the efficiency of an algorithm by analyzing how accurate an approximation to the affected region it identifies and on the overhead it incurs in doing this. In particular, we try to analyze the complexity of incremental algorithms and problems in terms of a parameter $\| \delta \|$ that measures the *size of the change in the input and output*. An incremental algorithm is said to be *bounded* if the time it takes to update the output depends only on the size of the *change* in the input and output (i.e., $\| \delta \|$), and not on the size of the *entire* current input. Otherwise, an incremental algorithm is said to be *unbounded*. A problem is said to be *bounded* (unbounded) if it has (does not have) a bounded incremental algorithm. The results established in this book illustrate a complexity hierarchy for incremental computation from this point of view. These results are summarized below.

We present efficient $O(\| \delta \| \log \| \delta \|)$ incremental algorithms for several shortest-path problems — the single-sink shortest-path problem, the all-pairs shortest-path problem, and a generalization of the single-sink shortest-path problem due to Knuth — establishing that these problems are all polynomially bounded. These results show that it is possible in these problems to identify, without much of an overhead, exactly the part of the output that needs to be updated.

We present an $O(2^{\| \delta \|})$ incremental algorithm for the circuit value annotation problem, which matches a previous $\Omega(2^{\| \delta \|})$ lower bound for this problem. Consequently, this establishes that the circuit value annotation problem is an exponentially bounded problem. We also present experimental results that show that our algorithm, in spite of a worst-case complexity of $\Theta(2^{\|\delta\|})$, works well in practice, often identifying a very good approximation to the affected output with very little overhead.

We present lower bounds showing that a number of problems, including graph reachability, dataflow analysis, and algebraic path problems, are unbounded with respect to a model of computation called the *sparsely-aliasing pointer machine* model.

We present an $O(\|\delta\| \log n)$ incremental algorithm for the reachability problem in reducible flowgraphs, which identifies the affected output exactly, but with an $O(\log n)$ factor overhead.

We present an algorithm for maintaining the dominator tree of a reducible flowgraph, which identifies a reasonable approximation to the affected output and updates it efficiently.

Acknowledgements

This book is a slightly revised version of my Ph.D. thesis. I am greatly indebted to my thesis advisor Thomas Reps for his guidance, both personal and professional, both while I was a student and after my graduation, which, among other things, shaped the thesis and its form. I thank also

- Susan Horwitz and Marvin Solomon, for their patient reading of my thesis and their valuable suggestions which helped improve the thesis;
- Eric Bach and Richard Brualdi, the other members of my thesis committee;
- Prasoon Tiwari, for various useful discussions I had with him;
- Charles Fischer, for his excellent teaching;
- the members of the "PL" group, in particular, Paul Adams, Tom Ball, Sam Bates, Dave Binkley, Lorenz Huelsbergen, Phil Pfeiffer, Todd Proebsting, and Wuu Yang;
- IBM, for partially supporting me through a graduate fellowship while I was a student, and for its support subsequently;
- my friends, too numerous to list here, who made my stay at Madison, Wisconsin, a memorable one;

I especially thank my family, without whose support this would not have been possibe. I thank my parents, my wife, and my sister for their love and support over the years.

May 1996   G. Ramalingam

Table Of Contents

Chapter 1. Introduction 1

Chapter 2. On Incremental Algorithms and Their Complexity 7

   2.1. Incremental Algorithms: Why and Where? 7

   2.2. Why Incremental Graph Algorithms? 9

      2.2.1. The Reachability Problem 9

      2.2.2. The Shortest-Path Problem 10

      2.2.3. The Circuit Value Annotation Problem 10

      2.2.4. The Dominator Tree Problem 11

   2.3. Evaluating Incremental Algorithms: Why and How? 11

      2.3.1. Analytical Evaluation of Incremental Algorithms 12

         2.3.1.1. Asymptotic Analysis Versus Micro-analysis 13

         2.3.1.2. Alternatives to Worst-Case Analysis 13

         2.3.1.3. Alternatives to Input Size as Complexity Parameters 16

         2.3.1.4. On Classifying Incremental Computation Problems 20

      2.3.2. Experimental Evaluation of Incremental Algorithms 22

Chapter 3. Terminology and Notation 25

Chapter 4. Incremental Algorithms for Shortest-Path Problems 30

   4.1. The Dynamic Single-Sink Shortest-Path Problem 30

      4.1.1. Deletion of an Edge 31

      4.1.2. Insertion of an Edge 36

      4.1.3. Handling Edges of Non-Positive Length 38

      4.1.4. A Batch SSSP Algorithm 42

   4.2. The Dynamic All-Pairs Shortest-Path Problem 43

      4.2.1. Deletion of an Edge 45

      4.2.2. Insertion of an Edge 48

      4.2.3. Handling Edges of Non-Positive Length 51

   4.3. Related Work 51

Chapter 5. Generalizations of the Shortest-Path Problem 52

   5.1. An Overview of the Chapter 52

   5.2. Maximum Fixed Point Problems 55

5.3. Path Problems in Graphs 56

    5.3.1. Algebraic Path Problems 56

    5.3.2. Dataflow Analysis Problems 61

    5.3.3. The Relation between Dataflow Analysis and Algebraic Path Problems 62

5.4. Grammar Problems 63

    5.4.1. The Idea Behind The Problem 63

    5.4.2. The Problem Definition 66

    5.4.3. The Grammar Problem as a Maximum Fixed Point Problem 69

    5.4.4. The SSF Grammar Problem and the SWSF Fixed Point Problem 71

5.5. Related Work 77

Chapter 6. An Incremental Algorithm for a Generalization of the Shortest-Path Problem 78

6.1. Introduction 78

6.2. The Idea Behind the Algorithm 79

6.3. The Algorithm 85

6.4. An Improved Algorithm 87

6.5. Extensions to the Algorithm 92

    6.5.1. Answering Queries on Demand 92

    6.5.2. Maintaining Minimum Cost Derivations 93

    6.5.3. The All-Pairs Shortest-Path Problem 94

    6.5.4. Handling Edges of Non-Positive Length 94

6.6. Related Work 95

Chapter 7. Incremental Algorithms for the Circuit Value Annotation Problem 101

7.1. Introduction 101

7.2. The Change Propagation Strategy 103

7.3. The Iterative Evaluate-and-Expand Strategy 107

7.4. Breadth-First Expansion 110

    7.4.1. Bounded-Outdegree Circuits 110

    7.4.2. Handling Unbounded Outdegree 112

    7.4.3. Monotonic Circuits: A Special Case 114

7.4.4. A Tradeoff 115

7.5. The Weighted Circuit Value Annotation Problem 119

7.6. The Empirical Boundedness of Incremental Algorithms 124

7.7. Some Remarks 128

Chapter 8. Inherently Unbounded Incremental Computation
   Problems 130

8.1. Introduction 130

8.2. The Model of Computation 130

   8.2.1. Locally Persistent Algorithms 130

   8.2.2. The Cost of Elementary Operations 131

   8.2.3. Sparsely Aliasing Algorithms 133

8.3. The Unboundedness of Reachability 137

8.4. Other Unbounded Problems 142

   8.4.1. Unbounded Path Problems 142

   8.4.2. Unbounded Dataflow Analysis Problems 144

8.5. Some Remarks 146

Chapter 9. Incremental Algorithms for Reducible Flowgraphs 147

9.1. Introduction 147

9.2. Reachability, Domination, and Reducible Flowgraphs 147

9.3. The Dynamic Single-Source Reachability Problem 149

   9.3.1. Insertion of an Edge 153

   9.3.2. Deletion of an Edge 153

   9.3.3. Handling Irreducible Flowgraphs 153

9.4. The Dynamic Dominator Tree Problem 156

   9.4.1. Insertion of an Edge 158

   9.4.2. Deletion of an Edge 164

   9.4.3. Related Work 166

   9.4.4. Some Remarks 167

Chapter 10. Conclusions 169

Bibliography 172

Chapter 1
Introduction

> *Observe constantly that all things take place by change, and accustom thyself to consider that the nature of the Universe loves nothing so much as to change the things which are, and to make new things like them.*
>
> —Marcus Aurelius

The subject of this book is incremental computation—computation with dynamic or changing data.

A *batch* algorithm takes an input and computes an output that is some function of the input. Such algorithms are also called *off-line* algorithms. *Incremental computation*, in contrast, is concerned with updating the output as the input undergoes changes. An *incremental algorithm* for computing a function f takes as input the "batch input" x, the "batch output" $f(x)$, possibly some auxiliary information, and a description of the "change in the batch input", Δx. The algorithm computes the "new batch output" $f(x + \Delta x)$, where $x + \Delta x$ denotes the modified input, and updates the auxiliary information as necessary. (See Figure 1.1.) A batch algorithm for computing f can obviously be used in this situation—it is called a *start-over* algorithm in this context. But often this will not be the most efficient way of maintaining the output. For instance, in many applications, the "input" data x is some data structure, such as a tree, graph, or matrix, while the "output" of the application, namely $f(x)$, represents some "annotation" of the x data structure—a mapping from more primitive elements that make up x, e.g., graph vertices, to some space of values. It is often the case that small changes in the input cause correspondingly small changes in the output and it would be more efficient to compute the new output from the old output rather than to recompute the entire output from scratch. Incremental algorithms, also called *dynamic algorithms* or *on-line algorithms*, do exactly this.

Examples of programs that make good use of incremental computation include spreadsheets and word processors. WYSIWYG—what you see is what you get—word processors, common in the world of personal computing, usually display part of a formatted document, and, as the user makes changes to the input document, they continuously re-format the document and update the displayed page. In general, incremental computation is potentially useful in any context in which users "build" or "construct" some "object" gradually, and this object has to be repeatedly processed in some fashion as it is being built up. Some examples of such contexts are program development, document development, and computer-aided design.

Changes that necessitate re-computation are not always made by human users. A number of programs repeatedly modify some data and re-process the modified data in the course of their normal execution. The motivation for much of the previous work on *dynamic data structures*, in fact, originates from such situations. Such a situation also arises, for example, in an optimizing compiler that iteratively optimizes an input program, by applying a sequence of optimizing transformations, one after another. Each transformation changes the program in some fashion, and may make it necessary to reconstruct various internal representations of the program for subsequent use.

This book presents a collection of new results—upper bounds, lower bounds, and experimental results—for various incremental-computation problems. What is common to all these results is that the complexity of the algorithm or the problem is measured, in each case, not in terms of the input size, as is usually done. Instead, the complexity is measured in terms of an adaptive parameter that is a measure of the size of the *change* in the input and the output. Thus, we express the complexity not as a function of the size of the input but as a function of the size of the shaded regions in Figure 1.1.

A common way to evaluate the computational complexity of algorithms is to use asymptotic worst-case analysis and to express the cost of the computation as a function of the size of the input. However, for incremental algorithms, this kind of analysis is sometimes not very informative. For example, when the cost of the computation is expressed as a function of the size of the (current) input, the worst-case complexity of several incremental graph algorithms is no better than that of an algorithm that performs the computation from scratch [Che76, Zad84, Hoo87, Car88, Mar90]. In some cases (again with costs expressed as a function of the size of the input), it has even been possible to show a lower-bound result for the problem itself, demonstrating that *no* incremental algorithm, satisfying certain conditions[1], for the problem can, in the worst case, run in time asymptotically better than the time required to perform the computation from scratch [Spi75, Eve85, Ber90]. For these reasons, worst-case analysis *with costs expressed as a function of the size of the input* can sometimes fail to help in making comparisons between different incremental algorithms.

This book explores a different way to analyze the computational complexity of incremental-computation problems. Instead of analyzing their complexity in terms of the size of the *entire* current input, we concentrate on analyzing incremental algorithms in terms of the size of an "adaptive" parameter, denoted by CHANGED, that captures the *changes* in the input and the output. For the moment, we define CHANGED informally as CHANGED = Δ *input* + Δ *output*, where Δ *input* represents the changes in the input data and Δ *output* represents the differences between the old solution $f(input)$ and the new solution $f(input + \Delta input)$. The size of CHANGED following an input change δ will be denoted by $\| \delta \|$.

---

[1] Spira and Pan [Spi75], for example, assume that all comparisons are between functions on edge weights—the comparisons are not allowed to take advantage of the structure of the graph. There may be incremental algorithms outside the model assumed for the lower bound results that, in fact, perform better than the established lower bounds. Thus, while Spira and Pan show a lower bound of $\Omega(n)$ for updating the minimum spanning tree when an edge's length is modified, there exists an $O(\sqrt{n})$ algorithm for the same problem[Epp93]. (Here n denotes the number of nodes in the graph.)

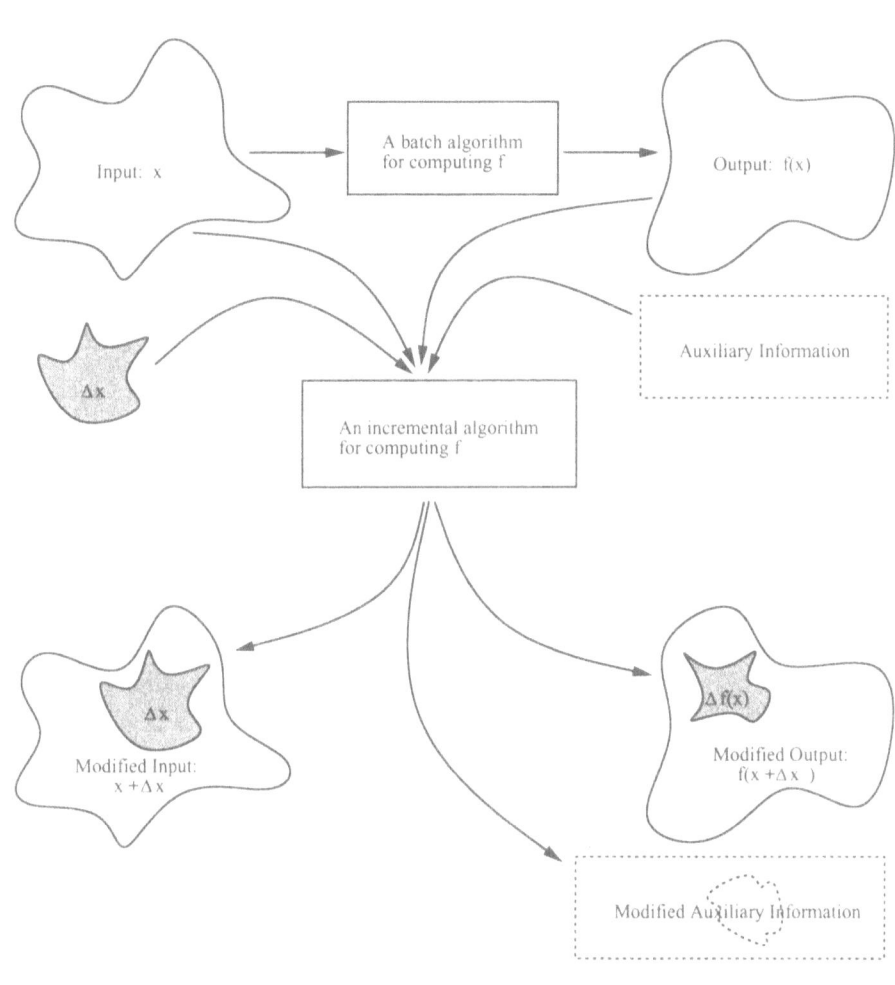

Figure 1.1. The above picture depicts the abstract problem of incremental computation. The shaded regions above denote the change in the input and the output. We denote the "size" of the change in the input and the output by $\|\delta\|$. The dotted lines around the auxiliary information signify that it is optional information maintained by the algorithm and that it can vary from incremental algorithm to incremental algorithm. The time taken by an incremental algorithm to process an input change will obviously depend on the cost of updating the auxiliary information. The parameter $\|\delta\|$ measures only the change in the input and the output and is, hence, a lower bound on the amount of work *any* incremental algorithm for the problem must do.

An incremental algorithm is said to be *bounded* if, for all input data-sets and for all changes that can be applied to an input data-set, the time it takes to update the output solution depends only on the size of the *change* in the input and output (*i.e.*, $\|\delta\|$), and not on the size of the *entire* current input. Otherwise, an incremental algorithm is said to be *unbounded*. A problem is said to be *bounded* (unbounded) if it has (does not have) a bounded incremental algorithm.

For certain problems, analyzing the complexity of an incremental algorithm in terms of the parameter $\|\delta\|$ provides a way to distinguish between different incremental algorithms where no such method was previously known. For instance, previous work has shown that, when the cost of the computation is expressed as a function of $|input|$, no incremental algorithm for the single-source shortest-path problem with positive edge lengths (SSSP>0) can (subject to certain restrictions) perform better than the best batch algorithm, in the worst case [Spi75, Eve85, Ber90]. In other words, with the usual way of analyzing incremental algorithms—worst-case analysis in terms of the size of the current input—no incremental shortest-path algorithm would appear to be any better than merely employing the best batch algorithm to recompute shortest paths from scratch! In contrast, we show in Chapter 4 that SSSP>0 has a bounded incremental algorithm that runs in time $O(\|\delta\| \log \|\delta\|)$ (whereas any batch algorithm for SSSP>0 will be an unbounded algorithm). This running time is within a log factor of the updating costs intrinsic to the problem.

A word of warning is appropriate here. Note that we do not suggest that the complexity of *all* incremental algorithms should be measured in terms of $\|\delta\|$. (On the contrary, the lower bounds established in this book show that even for some simple problems, the complexity of an incremental algorithm *cannot* be measured in terms of $\|\delta\|$ alone.)[2] The book shows that for *some* problems, $\|\delta\|$ turns out to be a natural and useful parameter for purposes of complexity analysis. Also note that this book considers a model of incremental computation where processing an input change requires computing the complete solution to the new input instance. Another, widely studied, model of incremental computation is the "update and query" model used in the study of "dynamic algorithms". In this model, a (dynamic) algorithm processes two types of requests: an "update" that changes the input in some way, and a "query" that asks for some part of the solution. The dynamic algorithm need not recompute the complete "solution" to the problem after an "update" to the input. Instead, it may choose to distribute the work that needs to be done between the update-processing and query-processing steps as appropriate.

The major contributions of this book are:

- Efficient incremental algorithms that can handle "unit" changes such as the insertion or deletion of a single edge for the single-sink/source shortest-path problem with positive length edges (SSSP>0) and the all-pairs shortest-path problem.

---

[2]The lower bounds established in this book show that some problems are unbounded; that is, no algorithm (of a certain class) exists whose asymptotic complexity can be expressed as a function of the parameter $\|\delta\|$ alone. While knowing that a problem is unbounded is informative in and of itself, it also means that if we want compare different algorithms for the problem, their complexities have to be expressed in terms of parameters other than (or in addition to) $\|\delta\|$.

- An efficient incremental algorithm that can handle arbitrary changes in the input for a generalization of the SSSP>0 problem.
- An incremental algorithm for the circuit value annotation problem that appears to be of practical utility.
- An incremental algorithm for maintaining reachability information in a flowgraph.
- An incremental algorithm for maintaining the dominator tree of a reducible flowgraph.
- Experimental results on the performance of three incremental algorithms for the circuit value annotation problem.
- Lower bound results for the incremental versions of various problems, such as graph reachability, dataflow analysis, and algebraic path problems.
- A complexity hierarchy for incremental computation that arises from the various results mentioned above.
- A survey of generalizations of the shortest-path problem along different dimensions, and an exploration of a new problem—the grammar problem—that combines these generalizations.

In the remainder of this chapter we present an overview of the organization of this book and discuss the above results in more detail.

In Chapter 2 we discuss the need for incremental computation in a number of applications and the relevance of incremental graph algorithms. We present an overview of the problems addressed in this book, discuss the difficulties involved in analyzing the computational complexity of incremental algorithms, motivate our approach to analyzing the complexity of incremental algorithms in terms of the parameter $\| \delta \|$, and present the complexity hierarchy of incremental computation that emerges when we adopt this approach.

In Chapter 3 we introduce the terminology and notation used in this book. In particular, we define the parameter $\| \delta \|$, and the concepts of bounded and unbounded incremental algorithms and problems.

In Chapter 4 we show that various shortest-path problems have efficient, polynomially-bounded, incremental algorithms. In particular, we present $O(\| \delta \| \log \| \delta \|)$ incremental algorithms for updating the solution to the *single-sink shortest-path problem with positive-length edges* (SSSP>0) and the *all-pairs shortest-path problem with positive-length edges* (APSP>0) after the insertion or deletion of an edge. We show that these algorithms can be adapted to work even in the presence of negative-length edges, as long as all cycles in the graph have a positive length. All these algorithms identify exactly the part of the output that needs to be updated and perform the update without much of an overhead.

Chapter 5 is a survey of various generalizations of the shortest-path problem. We review the definition of algebraic path problems and the lattice-theoretic formulation of dataflow analysis problems, and establish the equivalence of these problems. The equivalence of these two problems has been implicitly suggested in some of the previous work on these problem frameworks, but no formal relationship between the problems has been established to our knowledge. We discuss a generalization of the shortest-path problem due to Knuth. We show that these generalizations of the shortest-path problem, along different dimensions, can be naturally combined to yield an interesting problem, the *grammar problem* and study this problem. We show how some results relating dataflow analysis problems to maximal fixed point computation carry over to the grammar problem.

Chapter 6 presents an incremental algorithm for a generalization of the shortest-path problem. This generalized problem is a special case of the grammar problem. Like the algorithms presented in Chapter 4, the algorithm presented in this chapter identifies the part of the output that is affected exactly, without much overhead. In the case of the shortest-path problem, the results in this chapter show how the results in Chapter 4 can be generalized to the case of arbitrary changes in the input graph.

Chapter 7 presents exponentially bounded incremental algorithms for various versions of the circuit value annotation problem. The complexity of these algorithms matches a previously established lower bound for this problem. This lower bound suggests that it is not possible, in the circuit value annotation problem, to compute a good approximation to the affected output without much overhead. However, it is possible that the worst-case instances where this happens are rare. This chapter also presents experimental results on the performance of three different algorithms for the dynamic circuit value annotation problem, including a previously proposed algorithm, which show that, in practice, it might be possible to identify good approximations to the affected output without incurring much overhead.

Chapter 8 concerns lower bounds for incremental algorithms. The lower bounds are established with respect to a model of computation called the *sparsely-aliasing pointer machine* model. Graph reachability is shown to be unbounded with respect to this model of computation. Various algebraic path problems as well as dataflow analysis problems are shown to be unbounded by reducing the graph reachability problem to these problems.

Chapter 9 presents unbounded incremental algorithms for two related problems: the problem of maintaining reachability information for a reducible flowgraph, and the problem of maintaining the dominator tree of a reducible flowgraph. The incremental reachability algorithm can be adapted to work for arbitrary graphs, but it is particularly efficient for reducible flowgraphs with a time complexity of $O(\|\delta\| \log n)$. The reachability algorithm identifies the exact change necessary to the output, in the case of reducible flowgraphs, but incurs an overhead factor of $O(\log n)$ in doing this.

Chapter 10 concludes the book.

Chapter 2
On Incremental Algorithms and Their Complexity

> *In general, the time taken by an algorithm grows with the size of the input,*
> *so it is traditional to describe the running time of a program as a function of*
> *the size of its input. ... The best notion for input size depends on the problem*
> *being studied.*
>
> —Cormen, Leiserson, and Rivest, *Introduction to Algorithms*

The goal of this chapter is to present the motivation for the work described in this book, to describe the problems addressed in this book, to explain the reasons for the approach to complexity analysis adopted in this book, and to review related work briefly.

2.1. Incremental Algorithms: Why and Where?

Incremental computation is potentially useful in any context in which users "build" or "construct" some "object" gradually, and this object has to be repeatedly processed in some fashion as it is being built up. Such a context leads to the typical edit-process cycle where, in each iteration of the cycle, the users edit and modify the object, and then re-process it. We now give some examples of such situations and a brief overview of some work that has been done in these areas. (See [Ram93] for a categorized bibliography on incremental computation.)

Software Development

Programming, or software development, is one context where the situation described above is very common. Both programming-in-the-small and programming-in-the-large involve the edit-compile-execute cycle and the need for incremental processing has been long recognised in this area. The process of separate compilation and linking, in fact, grew out of this need. We can distinguish between two types of approaches to research in incremental processing in this area (and other areas as well). One approach centers around the use of an integrated system, which combines the editor and the processing tool. The user uses a special editor to edit the object he or she is creating—a program, in the case of software development—and the system performs the necessary re-processing as the user modifies the object. The second approach has been to concentrate on achieving incrementality in the various phases involved in compilation (or whatever processing one might wish to do). There has a considerable body of work on incremental algorithms for scanning, parsing, dataflow analysis, compilation, and linking. The generation of an incremental programming environment by integrating the various incremental phases is a separate problem, which has unfortunately not been addressed much.

Language Sensitive Editors and Attribute Updating. Language-sensitive editors incorporate knowledge about a specific programming language—its syntax and possibly some parts of its semantics as well. These are special editors meant for editing programs written in a particular programming language. Such editors can perform a semantic analysis of the (possibly incomplete) program being edited and provide immediate and useful feedback to the user about the syntactic and semantic errors in the program, as and when the user makes them. An example of a semantic error is a call to a function or procedure with the wrong number or type of arguments. Since these editors have to perform a semantic analysis of the program after each change the user makes to it, they have come to rely heavily on incremental static semantic analysis. Research in this area was initiated by Demers, Reps, and Teitelbaum (see [Dem81], [Rep82], [Rep83], and [Rep84]) who pioneered the use of an attribute-grammar based specification of a language and its semantics in automatically generating a editor for that specific language. There has been a lot of subsequent work in incremental attribute evaluation and in its applications, which are not restricted to static semantic analysis. The proceedings of recent conferences and workshops on attribute grammars [Der90, Alb91] are an excellent source of references to work in this area. Other work in this area includes: [Yeh83], [Joh83], [Joh82], [Joh85], [Jon86], [Rep86], [Kap86], [Hoo86] [Hoo87], [Fil87], [Yeh88], [Wal88], [Par88], [Kai89], [Tei90], [Jon90], [Zar90], [Fen90], [Vor90], [Alb90], [Pec90], [Hud91].

We list below some references to work on performing the various phases of compilation incrementally. See [Abm88] for a general discussion about incremental compilation.

Scanning and Parsing. [Ghe79], [Ghe80], [Weg80], [Jal82], [Agr83], [Kai85], [Ham88], [Bal88], [Jab88], [Hee90], [Jai90], [Bee91], [Wir92].

Dataflow Analysis. There has been a considerable amount of work in the area of incremental dataflow analysis. The recent dissertation of Marlowe [Mar89] is a good source for references in this area and for an overview of this area of research. Some of the other work in this area includes: [Ros81], [Ryd82], [Gho83], [Zad83], [Zad84], [Tan85], [Coo86], [Bur87], [Bur87a], [Car88], [Wan], [Ryd88], [Mar90], [Bur90], [Bur90a], [Ros90], [App91], [Pol92].

Compilation and Linking: [Fri83], [Fri84], [Fri84a], [Rei84], [Sch84], [For85], [Cro85], [Pol85], [Pol86], [Tic86], [Coo86], [Fuj88], [Tay88], [Sch88], [Tic88], [Lin89], [Ku89], [Gaf90], [Quo91], [Bur93], [Sha93].

Document Processing

Authors, either of small reports or large books, have a need, very similar to that of programmers, for incremental systems. While the edit-format-print-proof-read cycle is the analogue of the edit-compile-execute-debug cycle in programming, the WYSIWYG editors and document-formatters are the analogues of the language-sensitive editors in programming. The following are some references to work in this area: [Cha81], [Cha87], [Bro88], [Che88], [Che88a], [Che88b], [Har89], [Har91], [Mur92].

VLSI Design

Hardware design is accomplished today with the use of a number of CAD tools, which are often computation intensive. The iterative process of design has, consequently, engendered a need for incremental systems for hardware design. Some references to this area of research are: [Ous84], [Tay84], [Sco84], [Ous84a], [Car87], [Car89].

Some other contexts for incremental computation are:

Constraint Solving: [Bor79], [Kon84], [Van88], [Fre90], [Hen90], [Vor90a].

Reason Maintenance: [Doy79], [Doy79a], [Per84], [de86], [de86a], [de86b], [Sha88], [Smi88], [McA90].

Query Processing: [Shm84] [Shm90], [Wol91], [Man88].

2.2. Why Incremental Graph Algorithms?

This book deals with incremental graph algorithms. Why are incremental graph algorithms of interest? Graphs have become pervasive in the field of computer science because of their usefulness in modelling and formalizing problems in a number of areas. The use of graph algorithms is, consequently, very common in all kinds of tools. Graph-theoretic problems encountered in the areas of compilers and dataflow analysis, for instance, include the following: graph reachability, identifying the strongly connected components of a digraph, generating the SCC condensation of a graph, topologically sorting the vertices of a dag, constructing the dominator tree of a graph, and graph coloring [Aho86]. Graph problems of various kinds are also encountered in VLSI design.

In this section we briefly review the various problems addressed in this book, some of their applications, and the previous work on these problems. A more complete discussion of related work on these problems appears in the corresponding chapters later on.

2.2.1. The Reachability Problem

In the single-source reachability problem the input consists of a graph with a distinguished *source* vertex. The problem is to determine the set of all vertices in the graph that are reachable from the source vertex. In the all-pairs reachability problem, more commonly known as the transitive closure problem, one is interested in determining for every pair (u,v) of vertices in the graph if v is reachable from u.

Both the batch version and the dynamic version of the reachability problem have numerous applications and have attracted wide attention. See, for instance, [Ita86], [Pou88], [Ita88], [Yel], [Ber92], and [Mar92]. Note that the dynamic reachability problem is related to the problem of cycle-testing: the problem of checking if a graph remains acyclic as edges are inserted and deleted from it. The need for cycle-testing arises in deadlock detection and in unification. The dynamic reachability problem also arises in the areas of databases [Yan90], truth maintenance, belief revision [Smi88, McA90], and incremental rule processing [Wol91]. The dynamic reachability problem is closely linked to the incremental dataflow analysis problem, the problem of maintaining a strongly connected component condensation of a graph, and the dynamic domination problem. It also has connections to garbage collection.

2.2.2. The Shortest-Path Problem

The shortest-path problem, a combinatorial optimization problem, is a special case of the network flow problem. As Ahuja et al. [Ahu89] put it, "Shortest path problems are the most fundamental and also the most commonly encountered problems in the study of transportation and communication networks." Since shortest-path information can be used for routing in a communication network, the possibility of changes in the network, say due to a link failure, make the incremental shortest-path problem relevant to routing.

The applications of the shortest-path problem, however, are not restricted to the areas of transportation and communications networks. For instance, the shortest-path problem finds an interesting application in the problem of breaking paragraphs into lines (and the similar problem of determining appropriate page breaks for a document) in the context of document formatting. The following algorithm for determining the "optimum breakpoints", due to Knuth and Plass [Knu81], has been used in TeX. Given a paragraph that has to be formatted into a number of lines of length close to a given line-length, construct a graph in which the vertices represent the potential breakpoints—points where a line break can occur legally—in the paragraph. Add an edge between any two potentially successive breakpoints. The length of the edge denotes the "cost" of having all the text between these two breakpoints in a single line—for instance, if there is not much text between the two breakpoints, and the words would be spaced far apart, or if there is too much text and the line would look very cramped, then the cost would be high. One then looks for a shortest path from a vertex representing the beginning of the paragraph to a vertex representing the end of the paragraph.

A problem related to the shortest-path problem is that of determining the longest paths in an acyclic graph. This problem, the *critical path problem*, has applications in scheduling and circuit design, among other things.

Knuth [Knu77] proposed a generalization of the single-source shortest-path problem, which captures the flavor of dynamic programming, and discusses its applications. This problem is reviewed in Chapter 5.

A number of people have previously addressed various versions of the dynamic shortest-path problem, and their work is reviewed in Chapter 6.

2.2.3. The Circuit Value Annotation Problem

A *circuit* is a dag in which every vertex u is associated with a function F_u. Every vertex u corresponds to a value S_u that is to be computed by applying function F_u to the values computed at the predecessors of vertex u. The circuit value annotation problem is to compute the output value associated with each vertex. The dynamic version of the problem is to maintain consistent values at each vertex as the circuit undergoes changes [Par83, Rep83, Hoo87, Alp89, Alp90].

From a systems-building perspective, the dynamic circuit value annotation problem is important because it is at the heart of several important kinds of interactive systems, including spreadsheets [Bri79, Par83] and language-sensitive editors created from attribute-grammar specifications [Rep88]. The dynamic circuit value annotation problem is also of interest because the computation performed by an arbitrary program can be represented by a circuit and utilized in incremental execution of the same program [Hoo87]. Alphonse [Hoo92], a system for automatically generating efficient incremental systems from simple exhaustive imperative program specifications, makes use of incremental algorithms for the circuit value annotation problem.

Previous work on this problem includes [Par83], [Hoo87], [Alp90], and [Ram91]. Since the vertices in the circuit have to be evaluated in a topological order, this problem is also related to the problem of maintaining a topological ordering of the vertices in an acyclic graph, as the graph undergoes modifications such as the insertion and deletion of edges. However, it is not necessary to maintain a topological ordering of the vertices for the dynamic circuit value annotation problem. A dag is said to be *correctly prioritized* if every vertex u in the dag is assigned a priority, denoted by *priority* (u), such that if there is a path in the dag from vertex u to vertex v then *priority* (u) < *priority* (v). Alpern et al. [Alp90] outline an algorithm for maintaining a correct *prioritization* of the circuit, and use it for the dynamic circuit value annotation problem.

2.2.4. The Dominator Tree Problem

A vertex u, in a graph with a source vertex, is said to *dominate* a vertex v if all paths in the graph from the source vertex to v pass through u. The domination relation can be compactly represented by a *dominator tree*: a vertex u dominates a vertex v iff u is an ancestor of v in the dominator tree. The dominator tree plays an important role in several algorithms for program analysis and program optimization, and the need for updating the dominator tree of a flowgraph arises in various contexts such as incremental dataflow analysis. The only previous work that we are aware of on the dynamic dominator tree problem is [Car88a].

2.3. Evaluating Incremental Algorithms: Why and How?

Algorithms may be evaluated either analytically or experimentally. In this section we review some of the difficulties encountered in evaluating *incremental* algorithms using either of these methods and some of the previous work in this area. Similar observations on the difficulty of evaluating incremental algorithms appear in [Car88a, Ber92].

In evaluating an algorithm, one is interested in determining the resources required by the algorithm, including both the computational time required and the memory or space required. This section, like most work in this area, will concentrate primarily on the time requirements of incremental algorithms. It should be noted, however, that, in the context of incremental computation, the space requirements of algorithms *are* an important consideration, since some incremental algorithms can potentially end up storing and maintaining too much auxiliary information.

2.3.1. Analytical Evaluation of Incremental Algorithms

An analytic complexity measure of an *algorithm* provides us with an estimate of the resources required by that particular algorithm. An analytic complexity measure of a *problem* itself, on the other hand, describes "how difficult" that particular computational problem is. A comparison of two upper bounds, that is, the complexity measures of two algorithms for the same problem, can be useful in determining which of the two is a better algorithm. A comparison of a lower bound (that is, the complexity measure of a problem) with an upper bound (that is, the complexity measure of an algorithm for that problem) can be used to understand how well the particular algorithm does in relation to the intrinsic difficulty of the problem, and to figure out if there is much scope for improvement over the particular algorithm under consideration.

One of the first problems that one must come to grips with when dealing with algorithms for incremental-computation problems is that the criteria that one commonly uses to assess the performance of algorithms for batch-computation problems can be unsatisfactory. In particular, a common way to evaluate the time complexity of a batch algorithm is to use asymptotic analysis and to express the cost of the computation as a function of the size of the input; however, for incremental-computation problems, this kind of analysis can have several drawbacks:

- It may fail to distinguish between two different incremental algorithms for a problem, one of which is clearly superior to the other. (In many cases, it even fails to distinguish between an incremental algorithm and the batch start-over algorithm.)
- It can mislead one into believing that the batch start-over algorithm is optimal for a given incremental-computation problem.

For example, consider the problem of updating the attributes in a derivation tree of an attribute grammar after a tree modification. Both the incremental attribute-updating algorithm given in [Rep83] and the batch start-over algorithm have worst-case complexity of $O(|input|)$. Furthermore, because for some attributed trees certain modifications require every attribute in the tree to be given a new value, the incremental attribute-updating problem has a lower bound of $\Omega(|input|)$. Ordinarily one says that an algorithm whose asymptotic running time matches the lower bound for the problem is asymptotically optimal, from which one would conclude that the batch start-over algorithm is asymptotically optimal for the incremental attribute-updating problem.

The above example illustrates that asymptotic worst-case complexity measure, expressed as a function of the size of the input, can be inadequate in the case of incremental-computation problems. Several researchers have previously remarked on this inadequacy, and some have explored alternatives to this conventional complexity measure. These alternatives, which are reviewed below, include alternatives to asymptotic analysis, alternatives to worst-case analysis, parameters other than the size of the input, and various combinations of these alternatives.

A warning remark is appropriate before we consider these various alternative approaches. The above discussion should not be construed as suggesting that the conventional complexity measure fails completely in the case of incremental computation. There *are* problems for which incremental algorithms exist with a much better worst-case complexity measure than that of the batch start-over algorithm. An example is the problem of maintaining a minimum spanning tree of a graph. The standard batch algorithm for this problem runs in time $O(m \log n)$, while the incremental algorithm due to Frederickson [Fre85] updates the minimum spanning tree of a graph after the deletion or insertion of an edge in $O(\sqrt{m})$ time. Some other similar results include [Di89], [Epp92], and [Epp92a].

2.3.1.1. Asymptotic Analysis Versus Micro-analysis

In asymptotic analysis one ignores constant factors and expresses the complexity of algorithms using the big-O notation.[1] This has the advantages of simplifying the analysis and providing a machine-independent complexity measure. The possibility of using the constant factors to compare incremental algorithms with batch or start-over algorithms, when the two have the same asymptotic complexity measure, has been explored by Cheston [Che76]. In general, however, this may not be very useful, because incremental algorithms tend to have larger constant factors than the corresponding batch algorithms, especially when they have matching asymptotic worst-case complexity measures. Under such conditions, comparing constant factors would suggest that the batch algorithm is the better algorithm—a conclusion that is not necessarily warranted.

2.3.1.2. Alternatives to Worst-Case Analysis

Conventional worst-case complexity measure describes the time taken by the algorithm to process a worst-case input instance as a function of the input size. If one considers the set of all input instances of a given size, then the set of all worst-case input instances—those input instances that the algorithm takes the most time to process—normally forms some subset of this set. Worst-case complexity analysis has two advantages. The first is that the complexity measure provides a performance guarantee—the algorithm will process *any* input instance of a given size taking no more than the time guaranteed by the worst-case measure. The second advantage is that worst-case analysis is usually considerably simpler to perform than its alternatives listed below.

Worst-case analysis has been used with a large measure of success in analyzing batch algorithms. There is another reason for this success apart from the two

---

[1]In this book, we use standard notations for expressing the asymptotic behavior of functions. Asymptotic notation is discussed in most standard textbooks on algorithms, for example [Cor90].

advantages mentioned above: in many batch algorithms, the worst-case complexity measure matches the best-case complexity measure! In other words, every input instance is a worst-case input instance for these algorithms. For example, the straightforward matrix multiplication algorithm, for instance, takes time $\Theta(n^3)$ to multiply *any* two matrices of size $n \times n$. Similarly, a number of basic graph algorithms, such as those utilizing depth-first or breadth-first traversals, take time $\Theta(n+m)$ to process *any* graph with n vertices and m edges. The information provided by a worst-case analysis tends to become more approximate and less useful as worst-case input instances become a smaller subset of the set of all the input instances.

In many incremental algorithms and incremental-computation problems, worst-case input instances tend to be rare. Consider the problem of updating the solution to the single-source shortest-path problem after the deletion of a single edge. Let us compare the batch start-over algorithm with a simple incremental algorithm $\mathcal{A}$ that works as follows: the algorithm maintains a shortest-path tree for the graph; if the edge to be deleted is not in this shortest-path tree, the algorithm does nothing; if it is, then a batch algorithm is used to recompute a new shortest-path tree for the new graph. (We remark that most incremental algorithms for this problem do much better.) The worst case for this algorithm, and most incremental algorithms for this problem, occurs when all shortest paths in the currrent graph pass through a single edge, and this edge is deleted. The deletion of this single edge effectively changes the complete solution to the problem. This worst-case instance causes algorithm $\mathcal{A}$, as well as all known incremental algorithms, to perform as much work as the best batch algorithm, taking, say, $\Omega(m + n \log n)$ time.[2] The best case for this incremental algorithm, as well as other incremental algorithms, occurs when the deleted edge is not in any of the shortest paths. Algorithm $\mathcal{A}$ performs almost no work in this case, and the updating takes only a constant time. Now consider the following: A shortest-path tree for a graph with n vertices will have at most $n-1$ edges. If we assume that every edge in the graph is equally likely to be deleted, then algorithm $\mathcal{A}$ will do a non-trivial amount of work for only $n-1$ of the m possible edge deletions. But in the class of dense graphs $(n-1)/m$ tends to zero as n grows! Thus, in this case, the worst-case measure obviously does not convey as much information as it did in the examples of batch algorithms listed above.

An alternative to worst-case analysis is expected-case or average-case analysis. Average-case analysis can be very useful, but is typically much more difficult than worst-case analysis. Another problem with such an analysis is that it will not really provide a picture of the expected case unless the probability distribu-

---

[2] The complexity of a heap-based implementation of Dijkstra's shortest-path algorithm depends on the kind of heap used. If relaxed heaps are used, for instance, the complexity is $O(m + n \log n)$. Fredman and Willard's AF heaps improve the time complexity to $O(m + n \log n / \log \log n)$ [Fre90a].

tion used in the analysis reflects reality and the probability distribution for input changes, in the case of incremental computation, is very context-dependent. However, a simple, informal and approximate expected-case analysis, like the one in the previous paragraph, can be profitably used in demonstrating that an incremental algorithm is better than a batch algorithm. Examples of the use of expected-case analysis for incremental computation problems appear in [Che76, Eve85]. Pugh [Pug88] presents randomized data structures for use in incremental computation and an expected-case analysis of operations on these data structures.

Amortized analysis is another alternative to worst-case analysis that is applicable specifically to dynamic or on-line problems, where a sequence of operations is performed on some dynamic data structure. (See, for instance [Tar83], or [Cor90].) Amortized analysis concerns the *average* cost of an operation over a sequence of operations. Specifically, the amortized cost of an operation is defined to be the maximum over all possible sequences of operations of the average cost of an operation. Thus, the cost of any sequence of operations is guaranteed to be bounded by the product of the number of operations in the sequence and the amortized cost of an operation. In some situations, this bound on the cost of the sequence of operations is much better than the product of the number of operations and the worst-case cost of a single operation. In such cases, amortized analysis yields a better complexity measure than worst-case analysis.

Obviously, the amortized cost of an operation will be asymptotically better than the worst-case cost of an operation only if worst-case operations cannot appear too frequently in any sequence of operations. As Carroll observes,

> An algorithm with bad worst-case complexity will have good amortized complexity only if there is something about the problem being updated, or about the way in which we update it, or about the kinds of updates which we allow, that precludes pathological updates from happening frequently [Car88a].

Thus, although there are algorithms for dynamic problems that benefit from amortized-cost analysis, such as those of Even and Shiloach [Eve81], Reif [Rei87], and Ausiello *et al.* [Aus90], these benefits are obtained only by restricting the sequence of input modifications in some fashion. For example, the results of Even and Shiloach [Eve81] and Reif [Rei87] for dynamic graph connectivity hold only for a sequence of edge deletions, while the result of Ausiello *et al.* [Aus90] for maintaining shortest paths is applicable only in the case of a sequence of edge insertions. In the fully dynamic versions of these problems, where *both* edge insertions and edge deletions are allowed, "pathological" input changes can occur frequently in a sequence of input changes. Thus, the amortized-cost complexity of algorithms for the fully dynamic versions of these problems will not, in general, be better than their worst-case complexity. For instance, consider the worst-case scenario described above for the case of edge deletion in the dynamic shortest-path problem. The same edge can be repeatedly inserted and deleted. Consequently, the amortized complexity of an incremental algorithm for the fully dynamic shortest-path problem can be no better than the complexity of the batch algorithm.

2.3.1.3. Alternatives to Input Size as Complexity Parameters

We now turn our attention to the approach to complexity analysis used in this book. Complexity measures are expressed as a function of one or more parameters of the input—input size, for example. Let us refer to these parameters as *complexity parameters*. In the discussion of worst-case instances, in the previous section, we glossed over one fact: *the notion of worst-case input instances is not absolute but relative to the complexity parameter(s) one uses*. In fact, while worst-case input instances may be rare for some choice of the parameter(s), all input instances may be worst-case input instances for some other choice of the parameter(s), as we explain below.

The role of the complexity parameter(s) is as follows: The set of all possible input instances is partitioned into a number of classes, where each class C_n consists of the set of all input instances for which the parameter has a specific value n. Thus, if one used input size as the parameter, then the class C_n would consist of all the input instances of size n. An input instance in C_n is a worst-case instance for an algorithm iff the time the algorithm takes to process that input instance is roughly the maximum time the algorithm takes to process any input instance from the same class C_n. In worst-case analysis one attempts to express the time taken to process a worst-case input instance from a class C_n as a function of the parameter value n characterizing that class. Similarly, in average-case analysis one attempts to express the average time taken to process an input instance from a class C_n as a function of n.

If the parameter chosen is such that the time taken to process any input instance from a given class is roughly the same, then *all* input instances are worst-case input instances. In such a situation a worst-case complexity measure becomes an exact-case complexity measure. This argument captures the importance of the "parameter" one uses in measuring the complexity of an algorithm. It is preferable to choose the parameter so that the time taken to process an input instance depends only on the value of the parameter for that input instance. The reason worst-case analysis in terms of input size has been successful in analyzing batch algorithms is that the time taken to process an input instance correlates very well to the input size in batch computation.

The use of the parameters n, the number of vertices, and m, the number of edges, in analyzing graph algorithms illustrates the above point. Depth-first search in graphs, for instance, takes time $O(n+m)$. Expressed as a function of n, the complexity of depth-first search is $O(n^2)$. While depth-first search does take time $\Theta(n+m)$ for every graph with n vertices and m edges, it does not really take time $\Theta(n^2)$ for every graph with n vertices—it does so only when the number of edges in the graph is $\Theta(n^2)$. Thus, if we measure the complexity of depth-first search in terms of the parameters m and n every input instance is a worst-case input instance, while if we measure its complexity in terms of the parameter n, only dense graphs are worst-case input instances.

As another example, consider the single-source shortest-path problem with non-negative length edges. A straightforward implementation of Dijkstra's algorithm, as originally proposed by Dijkstra, runs in time $O(n^2+m) = O(n^2)$. An implementation of the same algorithm using ordinary heaps runs in time

$O((n+m)\log n)$, which is $O(n^2\log n)$ if one were to use only the parameter n. An implementation of the same algorithm using Fibonacci heaps runs in time $O(n\log n + m)$, which is $O(n^2)$ in terms of n alone. If one used only the parameter n, then one would be led to consider the straightforward implementation of Dijkstra's algorithm to be as good as the Fibonacci-heap based implementation and to consider it to be better than a simple-heap based implementation. However, both the heap-based implementations are arguably better than the straightforward implementation. If the complexity measures are expressed in terms of both the parameters n and m, then one is in a much better position to compare different graph algorithms.

It is not uncommon to find people turning to other parameters when a parameter like the input size proves to be inadequate in the above sense. Observe that an input instance for an incremental (graph) algorithm is an ordered pair (G, δ) consisting of a batch input instance G and a change δ to this instance. The approach used in this book is to measure the complexity of incremental algorithms in terms of a parameter $\|\delta\|_G$, which is a measure of the "size of the change in the input and the output" when change δ is applied to G. The parameter $\|\delta\|_G$, usually abbreviated to $\|\delta\|$, is formally defined in the next chapter.

In terms of the earlier discussion, the parameter $\|\delta\|$ induces a partition of the set of all ordered pairs (G, δ). Each class C_d in the partition consists of all pairs (G, δ) for which $\|\delta\|_G$ is d. Observe that there is no bound on the size of the graph G itself. Hence, it is possible that there exists no bound on the amount of time a given incremental algorithm takes to process an input change from class C_d. An incremental algorithm is said to be *bounded* if for every class C_d there is a bound on the amount of time the incremental algorithm takes to process an input instance from C_d. In other words, an incremental algorithm is said to be *bounded* if we can bound the time the algorithm takes to process a change δ to a graph G by some function of $\|\delta\|_G$. Otherwise, an incremental algorithm is said to be *unbounded*. A problem is said to be *bounded* (unbounded) if it has (does not have) a bounded incremental algorithm.[3]

The idea of using $\|\delta\|$ to measure the complexity of incremental algorithms is not new to this book. There have been a few previous papers in which incremental complexity has been measured in terms of the parameter $\|\delta\|$.

---

[3]There is another interesting way to think about boundedness. A bounded incremental algorithm will effectively update the solution to even an *infinite* problem instance G as long as the change in the input and output is finite. See Figure 2.1 for an example of an infinite input instance for the SSSP>0 problem, and a change to this infinite graph that causes only a finite change in the output. Even though such infinite graphs cannot be explicitly represented in a computer, the bounded incremental algorithm we present in Chapter 4 for the SSSP>0 problem is arguably an *effective procedure* for updating the solution even in this case.

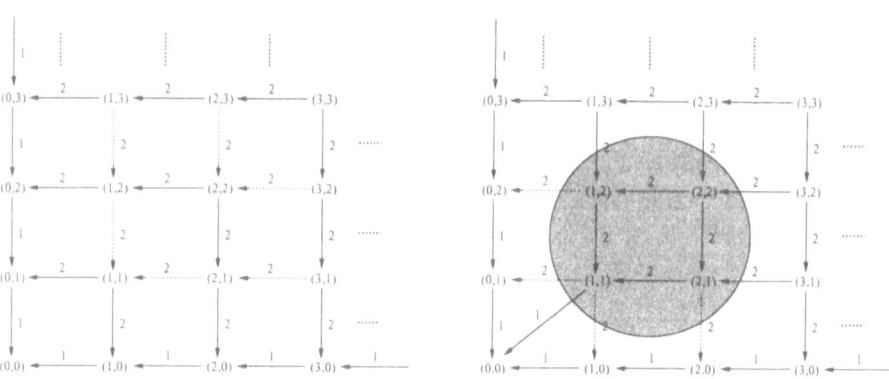

Figure 2.1. The figure on the left side represents an infinite graph whose vertices are points on the XY plane with non-negative integer coordinates. The origin $(0,0)$ is the sink vertex. There is an edge from vertex $(x+1, y)$ to vertex (x, y) and an edge from vertex $(x, y+1)$ to vertex (x, y), for every vertex (x, y). Edges on the X or Y axis have a length of 1, while all other edges have a length of 2. The solid edges in the graph indicate edges that are in some shortest paths, while the dotted edges are not in any of the shortest paths. The graph on the right is obtained by inserting an edge, of length 1, from the point $(1,1)$ to the point $(0,0)$. The shaded region indicates the affected vertices—vertices for which the length of the shortest path to the sink changes as a consequence of the edge insertion.

Attribute Updating

The first paper that we are aware of in which an incremental algorithm is analyzed in terms of $\|\delta\|$ is a paper by Reps [Rep82] (see also [Rep83] and [Rep84]).[4] The problem discussed in that paper is incremental attribute evaluation for noncircular attribute grammars—how to reevaluate the attributes of an attributed derivation tree after a restructuring operation (such as the replacement of a subtree) has been applied to the tree. The algorithm given is linear in $\|\delta\|$ and hence asymptotically optimal. Subsequently, other optimal algorithms were given for a variety of attribute-grammar subclasses, *e.g.*, absolutely noncircular grammars [Rep84] and ordered attribute grammars [Yeh83, Rep88].

In the case of the algorithms for noncircular attribute grammars and absolutely noncircular attribute grammars, the cost of an operation that moves the editing cursor in the tree is proportional to the length of the path along which the cursor is moved. (It is necessary to perform a unit-cost update to the auxiliary information used by the

---

[4]In these papers, the parameter $\|\delta\|$ is referred to as $|\text{AFFECTED}|$.

attribute updating algorithm at each vertex on the path along which the editing cursor is moved.) For ordered attribute grammars, however, a random-access movement of the editing cursor in the tree is a unit-cost operation.

There are also a variety of other attribute-updating algorithms described in the literature, including one that handles k simultaneous subtree replacements in an n-node tree and runs in amortized time $O((\| \delta \| + k) \cdot \log n)$ [Rep86], and another that permits unit-cost, random-access cursor motion for noncircular attribute grammars and runs in amortized time $O(\| \delta \| \cdot \sqrt{n})$ [Rep88a]. These algorithms have "hybrid" complexity measures, in the sense that the running time is a function of the size of the current input as well as $\| \delta \|$ (i.e., the running time is of the form $O(f(|input|, \| \delta \|))$.

Priority Ordering and the Circuit-Value Problem

A paper by Alpern et al. [Alp90] concerning the dynamic circuit-value problem and the problem of maintaining a priority ordering in a DAG presents results on the incremental complexity of both problems in terms of the parameter $\| \delta \|$.

(1) They showed that, with both edge insertions and deletions permitted, the problem of maintaining priorities in a DAG (as well as determining whether an edge insertion introduces any cycles) can be solved in time $O(\| \delta \|^2 \log \| \delta \|)$. In the case of unit changes, their algorithm runs in time $O(\| \delta \| \log \| \delta \|)$.

(2) They defined the concept of a locally persistent incremental algorithm (see Chapter 8), and showed that a lower-bound on any locally persistent algorithm for the dynamic circuit-value problem is $\Omega(2^{\| \delta \|})$.

(3) They gave an (unbounded) algorithm for the dynamic circuit-value problem that used their dynamic priority-ordering algorithm as a subroutine. In this algorithm, after a change to the graph, first priorities are updated; then, vertex re-evaluations are scheduled (via a worklist algorithm that uses a priority queue for the worklist). This algorithm runs in time

$$\| \delta_{PriorityOrdering} \|^2 \log \| \delta_{PriorityOrdering} \| + \| \delta_{CircuitValue} \| \log \| \delta_{CircuitValue} \| .$$

Since the quantity $\| \delta_{PriorityOrdering} \|$ is not bounded by any function of $\| \delta_{Circuit-Value} \|$, this algorithm for the dynamic circuit-value problem is unbounded.

More Recent Work

Subsequent to much of the work described in this book, several other related results have appeared. Berman [Ber92] presents several unboundedness results, with respect to the model of locally persistent algorithms, and also bounded incremental algorithms for reachability (and transitive closure) in acyclic graphs. Wiren [Wir93] presents a bounded incremental algorithm for natural language parsing. Frigioni et al. [Fri94] present incremental algorithms for the shortest path problem whose complexity they analyze in a similar way, by taking the change in input and output into account.

A Complexity Hierarchy

The results presented in this book, together with the previous results mentioned above, illustrate a complexity hierarchy that exists for incremental computation problems when their complexity is measured in terms of the parameter $\|\delta\|$. In particular, these results separate the classes of polynomially bounded problems, inherently exponentially bounded problems, and unbounded problems. The computational-complexity hierarchy for dynamic problems is depicted in Figure 2.2. This hierarchy exists with respect to a particular model of computation, which is discussed in Chapter 8.

2.3.1.4. On Classifying Incremental Computation Problems
In this section we look at related work on the problem of classifying incremental-computation problems. Each classification of incremental-computation problems provides, among other things, its own answers to the following questions: When can one

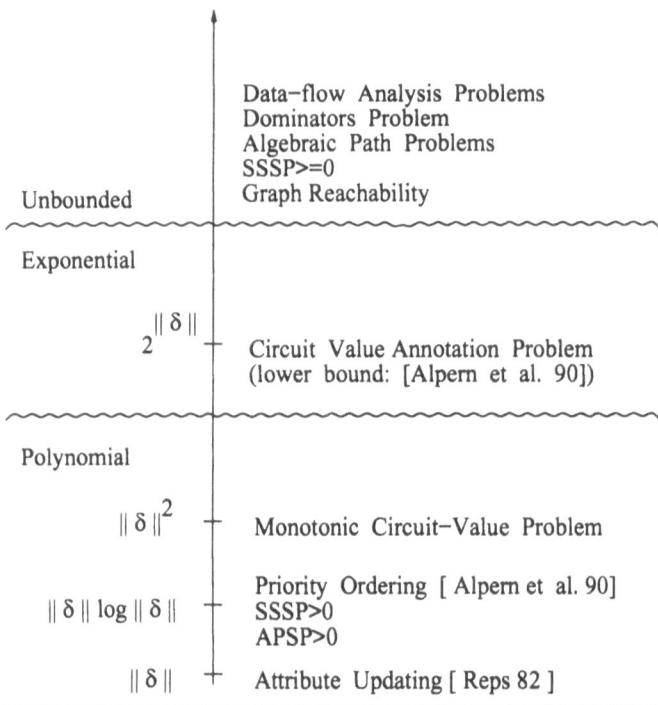

Figure 2.2. The results established in this book, along with a couple of previous results (indicated above by associated citations), illustrate a computational-complexity hierarchy for dynamic problems that exists when problems are classified according to their incremental complexity in terms of the parameter $\|\delta\|$ with respect to a model of computation called locally persistent algorithms.

say that an incremental-computation problem has a good algorithm? What does it mean for a problem to be "incrementalizable"?

The work outlined in this book attempts to use the notion of "boundedness" to distinguish problems that have good incremental solutions from those that do not. Arguably, incremental algorithms whose complexity is a low-order polynomial in $\| \delta \|$ are "good". Consequently, problems for which polynomially-bounded incremental algorithms exist can be said to have good incremental solutions. However, we do not mean to claim or suggest that unbounded incremental algorithms are necessarily poor incremental algorithms. For instance, we present in Chapter 9 an algorithm for maintaining reachability in a reducible flowgraph that runs in time $\Theta(\| \delta \| \log n)$. This is technically an unbounded algorithm, but it is, asymptotically, better than the batch algorithm, since its dependence on the size of the graph is only logarithmic.

One of the drawbacks with the boundedness approach is that it seems to be of limited applicability. For instance, all decision problems produce only a single bit output. Consequently, the concept of the "size of the change in the output" is not very useful in such cases.

A recent complexity-theoretic study of incremental computation [Sai93], explores *incr*-POLYLOGTIME, the class of problems whose dynamic versions are solvable in poly-logarithmic time (that is, time poly-logarithmic in input size), and *incr*-POLYLOGSPACE, the class of problems whose dynamic versions can be solved with poly-logarithmic work space. They suggest that these complexity classes capture the "intuitive notion of incremental efficiency"[5]. While these complexity classes are interesting enough that they merit study, it is perhaps not justifiable to consider only problems in this class to be problems with a good incremental algorithm. We have seen in earlier sections that problems exist for which any incremental algorithm, in the worst case, must do as much work as the best batch algorithm. Hoping for a poly-logarithmic incremental algorithm may often be too ambitious.

A different approach to the problem of classifying dynamic problems was proposed in a paper by Berman, Paull, and Ryder [Ber90]. (See also [Ber92].) They classify dynamic problems through the notion of an *incremental relative lower bound* (IRLB). A particular problem has an IRLB of $1/f(n)$ if it is possible to obtain any input instance of size n by making at most $f(n)$ unit changes to a "trivial input instance". Informally, a "trivial input instance" is an input instance for which the solution is easily obtained; see [Ber90] for a formal definition. Thus, if a problem has an IRLB of $1/f(n)$, then it is possible to compute the solution an instance of the batch problem of size n by starting with the solution to a trivial input instance and invoking an incremental algorithm for the problem at most $f(n)$ times. An IRLB pro-

---

[5]Observe the similarity here to parallel computation, where NC, the class of problems that have parallel algorithms that run in poly-logarithmic time using a polynomial number of processors, is usually considered to be the class of problems that are "parallelizable".

vides a lower bound for the incremental version of a problem in terms of a lower bound for the batch version of the problem. Thus, if a problem has an IRLB of $1/f(n)$ and the batch version of the problem has a lower bound of $g(n)$, then $g(n)/f(n)$ is a lower bound for the incremental version of the problem.

A limitation of the IRLB approach is that the lower bounds obtained apply only to incremental algorithms that satisfy some restrictions on the amount and kind of auxiliary information they use. In particular, the auxiliary information should not be difficult to compute initially—the algorithms cannot use much preprocessing time. For instance, the $O(\sqrt{n})$ dynamic minimum spanning tree algorithm described in [Epp92a, Epp93], which "contradicts" the lower bound of $\Omega(m/n)$ established using the IRLB technique, does not satisfy these restrictions.

Berman *et al.* discuss three classes of problems: those with an IRLB of 1, those with an IRLB of $1/n$, and those with an IRLB between $1/\sqrt{n}$ and $1/n$. In this classification scheme, the class of problems with an IRLB of 1 is the class with the poorest incremental behavior. For these problems, it is possible to show that a single modification, such as the insertion or deletion of a single edge in a graph, can change the problem to one whose solution shares nothing in common with the solution of the original problem (thereby reducing the batch problem to a "one-shot" dynamic problem).[6] Thus, in the worst case, an incremental algorithm for a problem with an IRLB of 1 cannot perform better than the best batch algorithm for the problem.

That a problem has an IRLB of 1 is certainly a property of interest (since the knowledge that there are modifications for which an incremental algorithm will perform no better than the best batch algorithm answers the question "How bad can things get?"), but it is not clear that problems with an IRLB of 1 should be considered to be problems with no good incremental algorithms. Some versions of the shortest-path problem have an IRLB of 1, yet from the point of view of boundedness have a good incremental algorithm.

2.3.2. Experimental Evaluation of Incremental Algorithms

Tarjan, in his Turing award lecture, says:

> Theoretical analysis of algorithms rests on sound foundations. This is not true of experimental analysis. We need a disciplined, systematic, scientific approach. Experimental analysis is in a way much harder than theoretical analysis because experimental analysis requires the writing of actual programs, and it is hard to avoid introducing bias through the coding process or through the choice of sample data.

We have seen in the previous section that theoretical analysis of *incremental* algorithms has its limitations and is not the final answer to questions concerning the per-

---

[6]The arguments that Berman, Paull, and Ryder use to establish relative lower bounds for various problems are similar to the ones used by Spira and Pan [Spi75] and Even and Gazit [Eve85] to establish that no incremental algorithm for the all-pairs shortest-path problem can do better in the worst case than the best batch algorithm for the problem.

formance of various incremental algorithms. For instance, none of the approaches to complexity analysis of incremental algorithms provides a satisfactory way of comparing the numerous algorithms that have been proposed for incremental dataflow analysis, and a similar situation exists with regard to a number of other problems that arise in practical systems. Consequently, there is a compelling motivation for experimental analysis of incremental algorithms

There have been relatively few papers in which the performance of an incremental algorithm has been evaluated from an experimental standpoint. The little work that does exist actually suggests that from a practical standpoint incremental algorithms that do not have "good" theoretical performance (according to criteria discussed in the previous section) can give satisfactory performance in real systems and work better than batch algorithms. For instance, Hoover presents evidence that his algorithm for the circuit-annotation problem performs well in practice [Hoo87]. Ryder, Landi, and Pande present evidence that the incremental dataflow analysis algorithm of Carroll and Ryder [Car88] performs well in practice [Ryd90]. Dionne reports excellent performance for some algorithms for the all-pairs shortest-path problem with positive edge weights [Dio78].

Experimental evaluation of incremental algorithms poses its own problems. Ryder *et al.* discuss some of the issues involved in evaluating an incremental algorithm [Ryd90]. One of the problems is the generation of suitable test data. While this can be difficult even for batch algorithms, it turns out to be even more difficult for incremental algorithms. For instance, input instances for dataflow analysis are control-flow graphs representing programs. Hence, it is quite easy to obtain typical input instances for the batch dataflow analysis problem. But for the incremental dataflow analysis problem one needs to generate typical sequences of modifications to the control-flow graph, and this can be difficult. A second problem is that incremental algorithms tend to be difficult to implement. If one is interested in comparing a number of incremental algorithms, implementing all of these algorithms is not an attractive proposition. (Programming systems and tools that can assist people in easily implementing, evaluating, and comparing incremental algorithms would be potentially useful in' this context.).

In the earlier discussion of analytic evaluation of incremental algorithms we observed the importance of the parameter used in describing the complexity measure. We noted that it was important to use a parameter to which the time complexity of the algorithm correlated well. This concern carries over to experimental evaluation of incremental algorithms as well. Experimentation yields data from which one can *empirically* determine a complexity measure of the algorithm. Consequently, one faces the same question—in terms of *what* parameter should one express the complexity of the algorithm?—in experimental evaluation of algorithms too.

The parameter $\| \delta \|$ turns out to be just as useful in the experimental evaluation of incremental algorithms as in the analytic evaluation of incremental algorithms. For instance, the graph in Figure 2.3 plots the time an incremental algorithm takes to process an input change as a function of $\| \delta \|$. Admittedly, the points do not fall on an ideal straight line. But, undeniably, the graph provides evidence that this

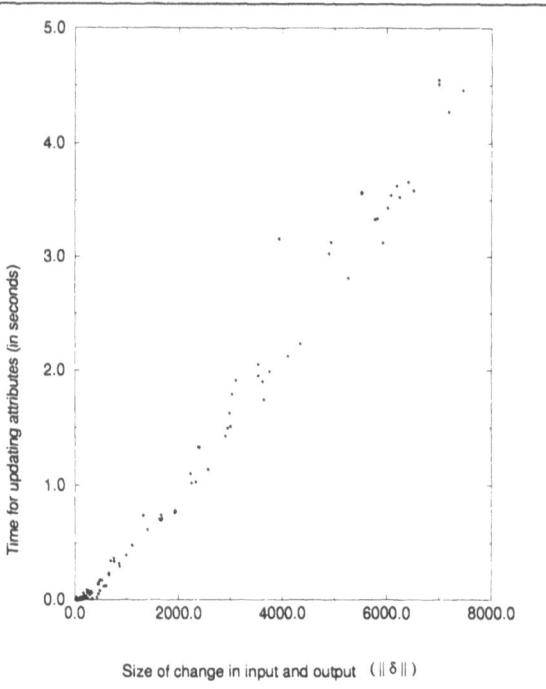

Figure 2.3. Performance results of one of the incremental algorithms for the circuit value annotation problem discussed in Chapter 7.

algorithm's complexity, in practice, is linear in $\| \delta \|$. This graph conveys more information than a simple plot of the "time taken to process an input change" as a function of "the input size" would have.

In Chapter 7 we present experimental results on the performance of three incremental algorithms for the circuit value annotation problem in the context of incremental attribute evaluation in a Pascal editor, illustrating the above point further.

Chapter 3
Terminology and Notation

> *Usually mathematicians avoid long theorems by the alternative device of long definitions ... this is more economical since one definition abbreviates many theorems. Even so, the definitions take up enormous space in 'rigorous' expositions ... the definition of 'ordinary polyhedra' in the 1962 edition of the Encyclopaedia Britannica fills 45 lines.*
> —I. Lakatos, *Proofs and Refutations: The Logic of Mathematical Discovery*

In this chapter we introduce the terminology and notation used in this book.

A *directed graph* $G = (V(G), E(G))$ consists of a set of *vertices* $V(G)$ and a set of *edges* $E(G)$, where $E(G) \subseteq V(G) \times V(G)$. An edge $(b,c) \in E(G)$, where $b, c \in V(G)$, is said to be directed from b to c, and will be more mnemonically denoted by $b \rightarrow c$. We say that b is the *source* of the edge, that c is the *target*, that b is a *predecessor* of c, and that c is a *successor* of b. A vertex b is said to be *adjacent* to a vertex c if b is a successor or predecessor of c. The set of all successors of a vertex a in G is denoted by $Succ_G(a)$, while the set of all predecessors of a in G is denoted by $Pred_G(a)$. If K is a set of vertices, then $Succ_G(K)$ denotes $\bigcup_{a \in K} Succ_G(a)$, and $Pred_G(K)$ is similarly defined. Given a set K of vertices in a graph G, the *neighborhood* of K, denoted by $N_G(K)$, is defined be the set of all vertices that are in K or are adjacent to some vertex in K: $N_G(K) = K \cup Succ_G(K) \cup Pred_G(K)$. The set $N_G^i(K)$ is defined inductively to be $N_G(N_G^{i-1}(K))$, where $N_G^0(K) = K$. Let $<F>$ denote the *subgraph induced* by a set of vertices F—that is, $<F> = (F, E(G) \cap (F \times F))$.

The *indegree* of a vertex u, denoted by $indegree_G(u)$, is the number of edges with u as the target, while the *outdegree* of a vertex u, denoted by $outdegree_G(u)$, is the number of edges with u as the source. The *degree* of a vertex is the sum of its indegree and outdegree. The subscript G will be dropped from the notation introduced above if no confusion is likely. The degree of a graph is defined to be the maximum degree of any vertex in the graph. The indegree and outdegree of a graph are similarly defined.

For any set of vertices K, we will denote the cardinality of K by both $|K|$ and V_K. We will denote the number of edges going out of vertices in K by $E_K^{\rightarrow}$: thus, $E_K^{\rightarrow} = \sum_{u \in K} outdegree(u)$. We similarly define $E_K^{\leftarrow}$ to be the number of edges coming into K, and we define E_K to be the number of edges that have at least one endpoint in K.

We now formulate a notion of the "size of the change in input and output" that is applicable to the class of graph problems in which the input consists of a graph G, and possibly some information (such as a real value) associated with each vertex or edge of the graph, and the output consists of a value $S_G(u)$ for each vertex u of the graph G. For instance, in the case of the single-sink shortest-path problem, $S_G(u)$ is the length of the shortest path from vertex u to a distinguished vertex, denoted by $sink(G)$. Thus, each vertex and/or edge in the graph may have an associated *input* value, and each vertex in the graph has an associated *output* value.

Before we formally define the parameter $\|\delta\|$ let us consider an example that will motivate the definition. In the circuit value annotation problem, which was introduced in the previous chapter, the output value for each vertex in the input dag is defined as some function of the values computed at its predecessor vertices. Let us view an input change as changing the function associated with one or more vertices—let us call these vertices the *modified* vertices. As a result of this change, the output value associated with some vertices (which may or may not be modified vertices) will change—let us call these vertices *affected* vertices. Obviously, any incremental algorithm for this problem will have to do at least work proportional to the number of modified and affected vertices. Can we have an incremental algorithm that does *no more* work than this?

This goal is too ambitious. If the output value of a vertex changes, then any successor of this vertex is a *potentially affected* vertex. Any incremental algorithm, in general, will need to examine such vertices to determine if they are affected vertices. In other words, the set of vertices that *must* be re-evaluated following an input change—let us call this set *MustEvaluate*—includes not only modified and affected vertices but also the successors of affected vertices. In general, evaluating a vertex's value will require examination of the value of each of its predecessors. This suggests that any incremental algorithm for this problem must, in fact, take time $\Omega(V_{MustEvaluate} + E^{\leftarrow}_{MustEvaluate})$ to process an input change. Consequently, we would like to define $\|\delta\|$ to be $V_{MustEvaluate} + E^{\leftarrow}_{MustEvaluate}$.

The above discussion is relevant to all the problems we consider in this book, since, in each case, we can express the output value of a vertex as some function of the output values of its neighbouring vertices. (The other problems differ from the circuit value annotation problem in that the input graphs may contain cycles in these other problems. Thus, while the circuit value annotation problem is the problem of computing the fixed point of a non-recursive collection of equations, the other problems require the computation of some particular fixed point of a possibly recursive collection of equations.) However, in a problem like the single-sink shortest-path problem, the output value of a vertex depends on the output value of its *successors* (as opposed to the circuit value annotation problem where the output value of a vertex depends on the output value of its predecessors). Thus, we need to exchange the roles of predecessors and successors in the above discussion to apply it to the single-sink shortest-path problem. Just to simplify matters, we will somewhat generously define $\|\delta\|$ in terms of the number of neighbours of vertices (rather than the number of successors or predecessors) so that the definition is uniformly applicable to all these problems.

The above discussion motivates the following terminology. If K is a set of vertices in a graph G, then the *extended size* of K (of order 1), denoted by $\|K\|_{1,G}$ or just $\|K\|$, is defined to be $V_K + E_K$. In other words, $\|K\|$ is the sum of the number of vertices in K and the number of edges with an endpoint in K. Similarly, we define $\|K\|^{\leftarrow}$ to be $V_K + E^{\leftarrow}_K$ and we define $\|K\|^{\rightarrow}$ to be $V_K + E^{\rightarrow}_K$. The extended size of K of order i, denoted by $\|K\|_{i,G}$ or just $\|K\|_i$, is defined to be $V_{N^{i-1}(K)} + E_{N^{i-1}(K)}$—in

other words, it is the extended size of $N^{i-1}(K)$. In this book, we are only concerned
with extended sizes of order 1 and order 2.

We now turn to the problem of defining the "size" of the change in the input
and output.

We denote by $G+\delta$ the graph obtained by making a change δ to graph G. A
vertex u in G or $G+\delta$ is said to have been *modified* by δ if δ inserted or deleted u, or
modified the input value associated with u, or inserted or deleted some edge incident
on u, or modified the information associated with some edge incident on u.[1] The set
of all modified vertices in $G+\delta$ will be denoted by $\text{MODIFIED}_{G,\delta}$. Note that this set
captures the change in the input. A vertex in $G+\delta$ is said to be an *affected* vertex
either if it is a newly inserted vertex or if its output value in $G+\delta$ is different from its
output value in G. Let $\text{AFFECTED}_{G,\delta}$ denote the set of all affected vertices in $G+\delta$.
This set captures the change in the output. We define $\text{CHANGED}_{G,\delta}$ to be
$\text{MODIFIED}_{G,\delta} \cup \text{AFFECTED}_{G,\delta}$. This set, which we occasionally abbreviate further
to just δ, captures the change in the input and output. The subscripts of the various
terms defined above will be dropped if no confusion is likely.

There is a small complication in considering the extended size of the set of
modified vertices. Deleted vertices and edges occur only in G, while inserted vertices
and edges occur only in $G+\delta$. Hence, we use the union of the two graphs G and $G+\delta$,
denoted by $\tilde{G}$, in measuring extended sizes. We use $\| \text{MODIFIED} \|_{i,\tilde{G}}$ as a measure
of the size of the change in input, $\| \text{AFFECTED} \|_{i,\tilde{G}}$ as a measure of the size of the
change in output, and $\| \text{CHANGED} \|_{i,\tilde{G}}$, which we abbreviate to $\| \delta \|_i$ is a measure
of the size of the change in the input and output. An omitted subscript i implies a
value of 1.

In summary, we use both V_δ and $|\delta|$ to denote the number of vertices that are
modified or affected, and E_δ to denote the number of edges that have at least one end-
point that is modified or affected, and $\| \delta \|$ to denote $V_\delta + E_\delta$.

There are a couple of generalizations necessary to the above definition of
$\| \delta \|$ under some conditions. For some problems the output may not be unique—
there may be multiple satisfactory outputs for a given input. An example is the problem of prioritizing a dag, where we are interested in assigning a priority to each vertex in the graph such that $priority(u) < priority(v)$ if there is path from u to v in the
dag. Obviously, there can be many correct prioritizations of a dag. In such problems,
the set of vertices whose output values have to be changed, following an input
change, is not uniquely defined. In other words, the notion of affected vertices is not
precisely defined. Corresponding to each possible new solution there is a set of
affected vertices, and a corresponding value for $\| \delta \|$. For such problems, $\| \delta \|$ is

---

[1]Thus, if an *edge* is modified we consider both the endpoints of the edge to be modified ver-
tices. Again, this simplification leads to a somewhat generous definition of $\| \delta \|$. When an
edge is modified, in some problems, like the circuit value annotation problem, it would be appropriate to consider the target of the edge to be modified, while for some problems like SSSP,
it would be appropriate to consider the source of the edge to be modified.

defined to be the *minimum* extended size of the change in the input and output over all possible new solutions [Alp90].

Another generalization becomes necessary when one considers problems where the output computed for a vertex is not *atomic* but *structured*. That is, the output for a vertex consists not of a single value, but of multiple values. We will define this generalization later in Chapter 4, when we discuss the all-pairs shortest-path problem, where one is interested in computing a vector of values for each vertex.

An incremental algorithm for a problem P takes as input a graph G, the solution to graph G, possibly some auxiliary information, and input change δ. The algorithm computes the solution for the new graph $G+\delta$ and updates the auxiliary information as necessary. The time taken to perform this update step may depend on G, δ, and the auxiliary information. An incremental algorithm is said to be a *bounded algorithm* if, for a fixed value of i, we can express the time taken for the update step entirely as a function of the parameter $\| \delta \|_{i,G}$ (as opposed to other parameters, such as $|V(G)|$). (The cost of updating really depends on the model of computation—it depends on the costs assigned to elementary operations. We use a form of the uniform cost measure [Tar83] under which every elementary operation is assigned unit cost, subject to restrictions on the size of the operands allowed. This cost measure is discussed in greater detail later on, in Chapter 8.) It is said to be an *unbounded algorithm* if its running time can be arbitrarily large for fixed $\| \delta \|_{i,G}$.

A dynamic problem is said to be a *bounded problem* with respect to a model of computation if it has a bounded algorithm within that particular model of computation. Similarly, a dynamic problem is said to be an *unbounded problem* with respect to a model of computation if it has no bounded algorithm within that model of computation.

While the above definition of boundedness is applicable for most of the problems we discuss in this book, there are some problems where it needs to be generalized. Let us return to a point raised earlier, in the discussion of the circuit value annotation problem. The motivation for considering the extended size of order 2 comes from the following two points: it is necessary to evaluate vertices that are one step away from affected vertices, and evaluating a vertex requires examining vertices that are a step away from the evaluated vertex. An alternative, more general, approach to using the extended size of order 2 is to assume that the cost of evaluating a function at a vertex u is given by some known constant $cost(u)$. For any set X of vertices, define $cost(X)$ to be the sum of the costs of all vertices in X. We earlier noted that any incremental algorithm for the circuit value annotation problem would have to evaluate all vertices in CHANGED $\cup$ *Succ* (AFFECTED). We define $C_{G,\delta}$, abbreviated to C_δ, to be $cost$(CHANGED $\cup$ *Succ* (AFFECTED)). Thus, C_δ is the minimum amount of work an algorithm would have to spend on evaluating functions during an update. An incremental algorithm is said to be a *bounded scheduling cost algorithm* if we can bound the time taken for the update step by a function of the parameters $\| \delta \|_G$ and $C_{G,\delta}$. We will similarly denote the maximum cost of evaluating any vertex in CHANGED $\cup$ *Succ* (AFFECTED) by M_δ.

By a *unit change* we mean a change that modifies the information associated with a single vertex or edge, or that adds or deletes a single vertex or edge. Other changes are said to be *non-unit changes*. If $f(u)$ is some information (either input or output) associated with a vertex u, we will refer to its value before and after the change by $f_{old}(u)$ and $f_{new}(u)$ respectively.

There are some comments worth making about the presentation of algorithms in the subsequent chapters. Most of these algorithms make use of sets and linked lists of various kinds. Operations on sets and lists are often specified at a high level. Some sophisticated implementations—*e.g.*, a doubly linked list—of these data structures may be required, in various cases, to make the relevant operations efficient. However, we do not specify the exact implementation to be used for the various sets and lists. Determining the implementation that is most appropriate in each case should not be difficult—it should, in fact, be within the capabilities of sophisticated compilers for high level languages such as SETL.

One final note: An alternative strategy for studying the computational complexity of incremental algorithms would have been to restrict the input instances to graphs with a fixed bound on indegree and outdegree, and to express incremental complexity as a function of the parameter $|\delta|_G = |\text{CHANGED}_{G,\delta}|$. Instead, we have chosen to work with problems on general graphs and express incremental complexity in terms of $\|\delta\|_{i,G} = \|\text{CHANGED}\|_{i,G+\delta}$. If one does restrict attention to families of bounded-degree graphs, all complexity bounds given in the paper of the form $O(f(\|\delta\|_{i,G}))$ can be restated as bounds of the form $O(f(k \cdot |\delta|_G))$, where the constant k depends on i and the maximum degree of the graph.

Chapter 4
Incremental Algorithms for Shortest-Path Problems

> *Access to the city was sharply restricted when the earthquake Tuesday*
> *broke the San Francisco-Oakland Bay Bridge ... As alternate routes across the*
> *bay, drivers had several main choices, and all of them posed problems. Taking*
> *a route to the north, and entering San Francisco over the Golden Gate Bridge*
> *meant going miles out of the way for the regular users of the Bay Bridge.*
> —New York Times (October 20, 1989)

This chapter presents polynomially bounded ($O(\|\delta\| \log \|\delta\|)$) incremental algorithms for processing unit changes for the single-source shortest-path problem and the all-pairs shortest-path problem with positive length edges. We also show how negative length edges can be handled as long as all cycles in the graph have a positive length. We describe an application of the incremental algorithm for the dynamic SSSP>0 problem to the batch SSSP problem on graphs that have a small number of negative edges (but no negative-length cycles). (The material presented in this chapter appears in [Ram96]).

4.1. The Dynamic Single-Sink Shortest-Path Problem

The input for SSSP>0 consists of a directed graph G with a distinguished vertex $sink(G)$. Every edge $u \rightarrow v$ in the graph has a real-valued length, which we denote by $length(u \rightarrow v)$. The length of a path is defined to be the sum of the lengths of the edges in the path. We are interested in computing $dist(u)$, the length of the shortest path from u to $sink(G)$, for every vertex u in the graph. If there is no path from a vertex u to $sink(G)$ then $dist(u)$ is defined to be infinity.

This section concerns the problem of updating the solution to an instance of the SSSP>0 problem after a unit change is made to the graph. The insertion or deletion of an isolated vertex can be processed trivially and will not be discussed here. We present algorithms for performing the update after a single edge is deleted from or inserted into the edge set of G. The operations of inserting an edge and decreasing the length of an edge are equivalent in the following sense: The insertion of an edge can be considered as the special case of an edge length being decreased from ∞ to a finite value, while the case of a decrease in an edge length can be considered as the insertion of a new edge parallel to the relevant edge. The operations of deleting an edge and increasing an edge length are similarly equivalent. Consequently, the algorithms we present here can be directly adapted for performing the update after a change in the length of an edge.

Proposition 4.1. *SSSP>0 has a bounded incremental algorithm. In particular, there exists an algorithm* DeleteEdge$_{SSSP>0}$ *that can process the deletion of an edge in time* $O(\|\delta\| + |\delta| \log |\delta|)$ *and there exists an algorithm* InsertEdge$_{SSSP>0}$ *that can process the insertion of an edge in time* $O(\|\delta\| + |\delta| \log |\delta|)$.

Though we have defined the incremental SSSP>0 problem to be that of maintaining the *lengths* of the shortest paths to the sink, the algorithms we present maintain the shortest paths as well. An edge in the graph is said to be an *SP* edge iff it occurs on some shortest path to the sink. Thus, an edge $u \rightarrow v$ is an *SP* edge iff $dist(u) = length(u \rightarrow v) + dist(v)$. A subgraph T of G is said to be a (single-sink) shortest-paths tree for the given graph G with sink $sink(G)$ if (i) T is a (directed) tree rooted at $sink(G)$, (ii) $V(T)$ is the set of all vertices that can reach $sink(G)$ in G, and (iii) every edge in T is an *SP* edge. Thus, for every vertex u in $V(T)$, the unique path in T from u to $sink(G)$ is a shortest path.

The set of all *SP* edges of the graph, which we denote by $SP(G)$, induces a subgraph of the given graph, which we call the shortest-paths subgraph. A graph and its shortest-paths subgraph are shown in Figure 4.1. We will occasionally denote the shortest-paths subgraph also by $SP(G)$. Note that a path from some vertex u to the sink vertex is a shortest path iff it occurs in $SP(G)$ (*i.e.*, iff all the edges in that path occur in $SP(G)$). Since all edges in the graph are assumed to have a positive length, any shortest path in the graph must be acyclic. Consequently, $SP(G)$ is a directed acyclic graph (dag). As we will see later, this is what enables us to process input changes in a bounded fashion. If zero length edges are allowed, then $SP(G)$ can have cycles, and the algorithms we present in this chapter will not work correctly in all instances.

Our incremental algorithm for SSSP>0 works by maintaining the shortest-path subgraph $SP(G)$. We will also find it useful to maintain the outdegree of each vertex u in the subgraph $SP(G)$.

4.1.1. Deletion of an Edge

The update algorithm for edge deletion is given as procedure DeleteEdge$_{SSSP>0}$ in Figure 4.2.

We will find it useful in the following discussion to introduce the concept of an *affected edge*. An *SP* edge $x \rightarrow y$ is said to be affected by the deletion of the edge $v \rightarrow w$ if there exists no path in the new graph from x to the sink that makes use of the edge $x \rightarrow y$ *and* has a length equal to $dist_{old}(x)$. It is easily seen that $x \rightarrow y$ is an affected *SP* edge iff y is an affected vertex. On the other hand, any vertex x other than v is an affected vertex iff all *SP* edges going out of x are affected edges. The vertex v itself is an affected vertex iff $v \rightarrow w$ is the only *SP* edge going out of vertex v.

The algorithm for updating the solution (and $SP(G)$) after the deletion of an edge works in two phases. The first phase (lines [4]–[14]) computes the set of all affected vertices and affected edges and removes the affected edges from $SP(G)$, while the second phase (lines [15]–[30]) computes the new output value for all the affected vertices and updates $SP(G)$ appropriately.

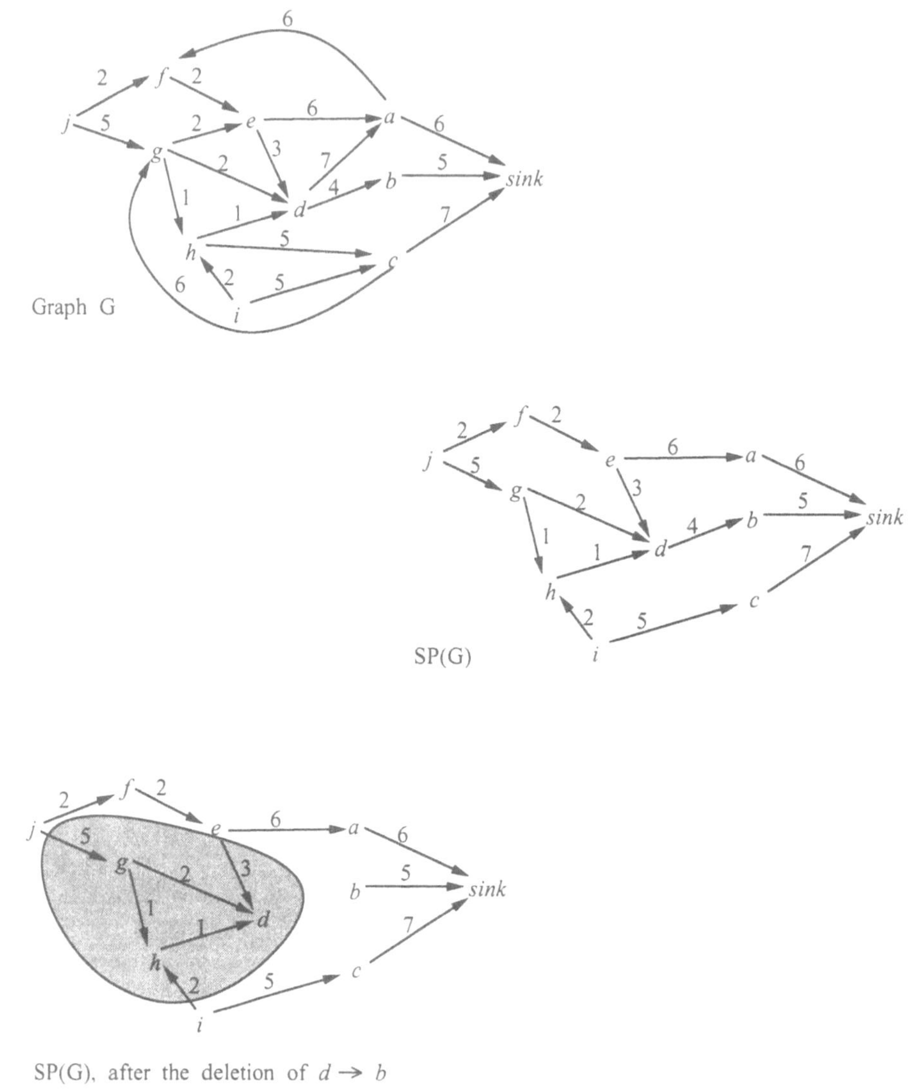

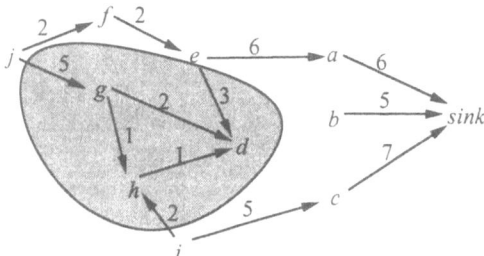

Graph G

SP(G)

SP(G), after the deletion of $d \rightarrow b$

Figure 4.1. A graph G and its shortest-paths subgraph $SP(G)$. The shaded region indicates the vertices and edges that are affected by the deletion of the edge $d \rightarrow b$.

---

procedure DeleteEdge$_{SSSP>0}$ $(G, v \rightarrow w)$
declare
        G: a directed graph;
        $v \rightarrow w$: an edge to be deleted from G
        WorkSet, AffectedVertices: sets of vertices;
        PriorityQueue: a heap of vertices
preconditions
        $SP(G)$ is the shortest-paths subgraph of G
        $\forall v \in V(G)$, $outdegree_{SP}(v)$ is the outdegree of v in the shortest-paths subgraph $SP(G)$
        $\forall v \in V(G)$, $dist(v)$ is the length of the shortest path from v to $sink(G)$
begin
[1]   **if** $v \rightarrow w \in SP(G)$ **then**
[2]   Remove edge $v \rightarrow w$ from $SP(G)$ and from $E(G)$ and decrement $outdegree_{SP}(v)$
[3]   **if** $outdegree_{SP}(v) = 0$ **then**
[4]   /* Phase 1: Identify the affected vertices and remove the affected edges from $SP(G)$ */
[5]   WorkSet := $\{ v \}$
[6]   AffectedVertices := $\emptyset$
[7]   **while** WorkSet $\neq \emptyset$ **do**
[8]   Select and remove a vertex u from WorkSet
[9]   Insert vertex u into AffectedVertices
[10]   **for** every vertex x such that $x \rightarrow u \in SP(G)$ **do**
[11]   Remove edge $x \rightarrow u$ from $SP(G)$ and decrement $outdegree_{SP}(x)$
[12]   **if** $outdegree_{SP}(x) = 0$ **then** Insert vertex x into WorkSet **fi**
[13]   **od**
[14]   **od**
[15]   /* Phase 2: Compute distances from affected vertices to $sink(G)$ and update $SP(G)$. */
[16]   PriorityQueue := $\emptyset$
[17]   **for** every vertex $a \in$ AffectedVertices **do**
[18] $dist(a) := \min (\{\ length(a \rightarrow b) + dist(b)\ |$
                      $a \rightarrow b \in E(G)$ and $b \notin$ AffectedVertices) $\} \cup \{ \infty \})$
[19]   **if** $dist(a) \neq \infty$ **then** InsertHeap(PriorityQueue, a, $dist(a)$) **fi**
[20]   **od**
[21]   **while** PriorityQueue $\neq \emptyset$ **do**
[22] $a :=$ FindAndDeleteMin(PriorityQueue)
[23]   **for** every vertex $b \in Succ(a)$ such that $length(a \rightarrow b) + dist(b) = dist(a)$ **do**
[24]   Insert edge $a \rightarrow b$ into $SP(G)$ and increment $outdegree_{SP}(a)$
[25]   **od**
[26]   **for** every vertex $c \in Pred(a)$ such that $length(c \rightarrow a) + dist(a) < dist(c)$ **do**
[27] $dist(c) := length(c \rightarrow a) + dist(a)$
[28]   AdjustHeap(PriorityQueue, c, $dist(c)$)
[29]   **od**
[30]   **od**
[31]   **fi**
[32]   **else** Remove edge $v \rightarrow w$ from $E(G)$
[33]   **fi**
end
postconditions
        $SP(G)$ is the shortest-paths subgraph of G
        $\forall v \in V(G)$, $outdegree_{SP}(v)$ is the outdegree of v in the shortest-paths subgraph $SP(G)$
        $\forall v \in V(G)$, $dist(v)$ is the length of the shortest path from v to $sink(G)$

Figure 4.2. An algorithm to update the SSSP>0 solution and $SP(G)$ after the deletion of an edge.

Phase 1: Identifying affected vertices

A vertex's *dist* value increases due to the deletion of edge $v \rightarrow w$ iff all shortest paths from the vertex to *sink*(G) make use of edge $v \rightarrow w$. In other words, if $SP(G)$ denotes the *SP* dag of the original graph, then the set of affected vertices is precisely the set of vertices that can reach the sink in $SP(G)$ but not in $SP(G) - \{v \rightarrow w\}$, the dag obtained by deleting edge $v \rightarrow w$ from $SP(G)$.

Consider the example shown in Figure 4.1. Consider the deletion of the edge $d \rightarrow b$ from the graph. To determine the set of affected vertices we remove $d \rightarrow b$ from $SP(G)$ and determine the set of vertices in $SP(G)$ that can no longer reach the sink vertex. The set of affected vertices and affected edges, in this case, is shown by the shaded region in the figure. Note that there exist shortest paths in G from vertices e, f, i, and j to the sink that pass through the deleted edge $d \rightarrow b$. However, these vertices are unaffected since there exist alternate shortest paths from these vertices to the sink that do not pass through edge $d \rightarrow b$. On the other hand, all shortest paths from vertices d, g, and h to the sink pass through edge $d \rightarrow b$, and, hence, these vertices are affected by the deletion of $d \rightarrow b$.

Thus, Phase 1 is essentially an incremental algorithm for the single-sink reachability problem in dags that updates the solution after the deletion of an edge. The algorithm is very similar to the topological sorting algorithm. It maintains a set of vertices (WorkSet) that have been identified as being affected but have not yet been processed. Initially v is added to this set if $v \rightarrow w$ is the only *SP* edge going out of v. The vertices in WorkSet are processed one by one. When a vertex u is processed, all *SP* edges coming into u are removed from $SP(G)$ since they are affected edges. During this process some vertices may be identified as being affected (because there no longer exists any *SP* edge going out of those vertices) and may be added to the workset.

We maintain $outdegree_{SP}(x)$, the number of *SP* edges going out of vertex x, so that the tests in lines [3] and [12] can be performed in constant time. We have not discussed how the subgraph $SP(G)$ is maintained. If $SP(G)$ is represented by maintaining (adjacency) lists at each vertex of all incoming and outgoing *SP* edges, then it is not necessary to maintain $outdegree_{SP}(x)$ separately, since $outdegree_{SP}(x)$ is zero iff the list of outgoing SP edges is empty. Alternatively, we can save storage by not maintaining $SP(G)$ explicitly. Given any edge $x \rightarrow y$, we can check if that edge is in $SP(G)$ in constant time, by checking if $dist(x) = length(x \rightarrow y) + dist(y)$. In this case, it is useful to maintain $outdegree_{SP}(x)$.

We now analyze the time complexity of Phase 1. The loop in lines [7]–[14] performs exactly $|\text{AFFECTED}|$ iterations, once for each affected vertex u. The iteration corresponding to vertex u takes time $O(|Pred(u)|)$. Consequently, the running time of Phase 1 is $O(\sum_{u \in \text{AFFECTED}} |Pred(u)|) = O(\|\text{AFFECTED}\|^{\leftarrow})$. If we maintain the *SP* dag explicitly, then the running time is linear in the extended size of AFFECTED in the *SP* dag, which can be less than the extended size of AFFECTED in the graph G itself.

Phase 2: Determining new distances for affected vertices and updating SP(G)

Phase 2 of DeleteEdge$_{SSSP>0}$ is an adaptation of Dijkstra's batch shortest-path algorithm that uses priority-first search [Sed83] to compute the new *dist* values for the affected vertices.

Consider Figure 4.3. Assume that for every vertex y in set A the length of the shortest path from y to the sink is known and is given by *dist*(y). We need to compute the length of the shortest path from x to the sink for every vertex x in the set of remaining vertices, B. Consider the graph obtained by "condensing" A to a new sink vertex: that is, we replace the set of vertices A by a new sink vertex s, and replace every edge $x \rightarrow y$ from a vertex x in B to a vertex y in A by an edge $x \rightarrow s$ of length

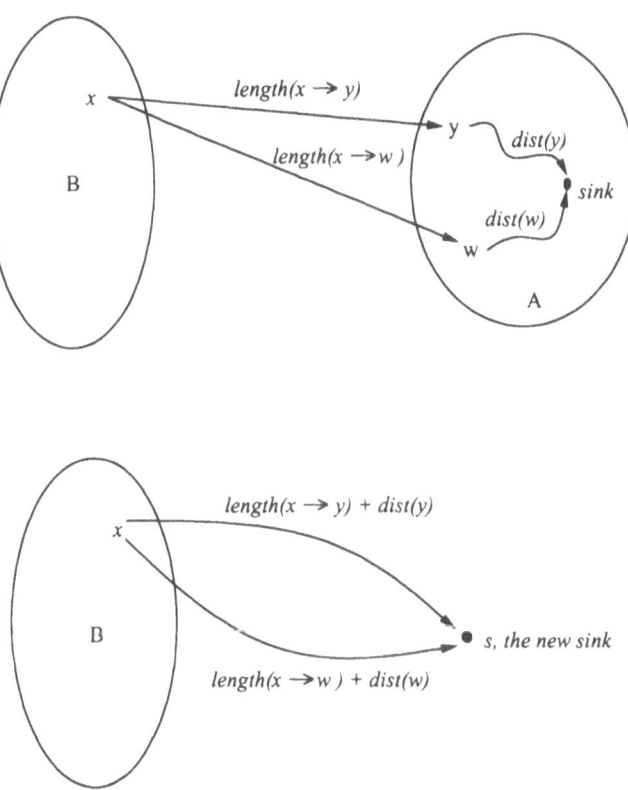

Figure 4.3. Phase 2 of DeleteEdge$_{SSSP>0}$. Let A be the set of unaffected vertices and let B be the set of affected vertices. The correct *dist* value is known for every vertex in A and the new *dist* value has to be computed for every vertex in B. This problem can be reduced to a *batch instance* of the SSSP>0 problem, namely the SSSP>0 problem for the graph obtained as follows: we take the subgraph induced by the set B of vertices, introduce a new sink vertex, and for every edge $x \rightarrow y$ from a vertex in B to a vertex outside B, we add an edge from x to the new sink vertex, with length *length* ($x \rightarrow y$) + *dist* (y).

$length(x \longrightarrow y) + dist(y)$. The given problem reduces to the SSSP problem for this reduced graph, which can be solved using Dijkstra's algorithm. Phase 2 of our algorithm works essentially this way.

A similar situation arises during the course of execution of Dijkstra's algorithm, where we know the *dist* values for a subset A of vertices, and the *dist* values have to be computed for the remaining vertices B. In relating the above situation in the incremental algorithm to the situation in the batch algorithm one should note the following point. During the course of execution of Dijkstra's algorithm we can guarantee that the correct *dist* value for every vertex in B is greater than or equal to the already computed *dist* value for any vertex in A. This is not necessarily true in the situation in the dynamic SSSP>0 problem.

Before we analyze the complexity of Phase 2, we explain the heap operations we make use of in the algorithm. The operation *InsertIntoHeap*(H, i, k) inserts an item i into heap H with a key k. The operation *FindAndDeleteMin*(H) returns the item in heap H that has the minimum key and deletes it from the heap. The operation *AdjustHeap*(H, i, k) inserts an item i into *Heap* with key k if i is not in *Heap*, and changes the key of item i in *Heap* to k if i is in *Heap*. In this algorithm, *AdjustHeap* either inserts an item into the heap, or decreases the key of an item in the heap.

The complexity of Phase 2 depends on the type of heap we use. We assume that PriorityQueue is implemented as a relaxed heap (see [Dri88]). Both insertion of an item into a relaxed heap and decreasing the key of an item in a relaxed heap cost $O(1)$ time, while finding and deleting the item with the minimum key costs $O(\log p)$ time, where p is the number of items in the heap.

The loop in lines [21]-[30] iterates at most |AFFECTED| times. An affected vertex a is processed in each iteration, but not all affected vertices may be processed. In particular, affected vertices that can no longer reach the sink vertex will not be processed. Each iteration takes $O(\|\{a\}\|)$ time for lines [23]-[29], and $O(\log |\text{AFFECTED}|)$ time for the heap operation in line [22]. Hence, the running time of Phase 2 is $O(\|\text{AFFECTED}\| + |\text{AFFECTED}| \log |\text{AFFECTED}|)$.

It follows from the bounds on the running time of Phase 1 and Phase 2 that the total running time of DeleteEdge$_{SSSP>0}$ is bounded by $O(\|\text{AFFECTED}\| + |\text{AFFECTED}| \log |\text{AFFECTED}|)$, which is $O(\|\delta\| + |\delta| \log |\delta|)$.

4.1.2. Insertion of an Edge

We now turn to the problem of updating distances and the set $SP(G)$ after an edge $v \longrightarrow w$ with length c is inserted into G. The algorithm for this problem, procedure InsertEdge$_{SSSP>0}$, is presented in Figure 4.5. The algorithm presented works correctly even if the length of the newly inserted edge is non-positive as long as all edges in the original graph have a positive length and the new edge does not introduce a cycle of negative length. This will be important in generalizing our incremental algorithm to handle edges of non-positive lengths.

The algorithm is based on the following characterization of the region of affected vertices, which enables the updating to be performed in a bounded fashion.

If the insertion of edge $v \rightarrow w$ causes u to be an affected vertex, then any new shortest path from u to $sink(G)$ must consist of a shortest path from u to v, followed by the edge $v \rightarrow w$, followed by a shortest path from w to $sink(G)$. In particular, a vertex u is affected iff $dist(u,v) + length(v \rightarrow w) + dist_{old}(w) < dist_{old}(u)$, where $dist(u,v)$ is the length of the shortest path from u to v in the new graph, and $dist_{old}$ refers to the lengths of the shortest paths to the sink in the graph before the insertion of the edge $v \rightarrow w$. The new $dist$ value for an affected vertex u is given by $dist(u,v) + length(v \rightarrow w) + dist_{old}(w)$.

Consider T, a single-sink shortest-path tree for the vertex v. Let x be any vertex, and let u be the parent of x in T. (See Figure 4.4.) If x is an affected vertex, then u must also be an affected vertex: otherwise, there must exist some shortest path P from u to $sink(G)$ that does not contain edge $v \rightarrow w$; the path consisting of the edge $x \rightarrow u$ followed by P is then a shortest path from x to $sink(G)$ that does not contain edge $v \rightarrow w$; hence, x cannot be an affected vertex, contradicting our assumption. In other words, any ancestor (in T) of an affected vertex must also be an affected vertex. The set of all affected vertices must, hence, form a connected subtree of T at the root of T.

The algorithm works by using an adaptation of Dijkstra's algorithm to construct the part of the tree T restricted to the affected vertices (the shaded part of T in Figure 4.4) in lines [3]-[6], [10], [11], and [16]-[19]. These lines differ from a straightforward implementation of Dijkstra's algorithm in the following way. When the vertex u is selected from PriorityQueue in line [11], its priority is nothing but $dist(u,v)$. In a normal implementation of Dijkstra's algorithm, every predecessor x of u would then be examined (as in the loop in lines [16]-[23]), and its priority in PriorityQueue would be adjusted if $length(x \rightarrow u) + dist(u,v)$ was less than the length of the shortest path found so far from x to v. Here, we instead adjust the priority of x or insert it into PriorityQueue *only if* $length(x \rightarrow u) + dist(u)$ is less than $dist(x)$: that

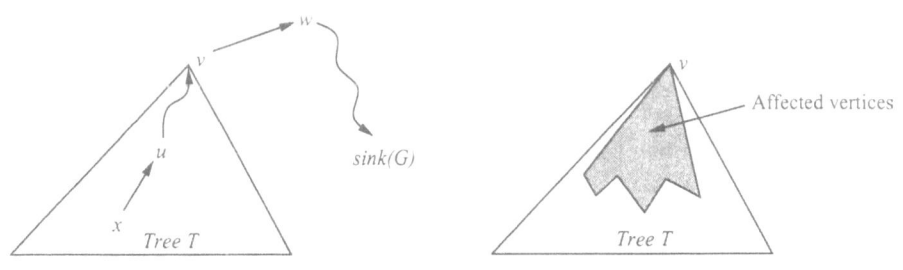

Figure 4.4. T is a shortest-path tree for sink v. If x is an affected vertex, then u, the parent of x in T, must also be an affected vertex. Hence, the set of all vertices affected by the insertion of the edge $v \rightarrow w$ forms a connected subtree at the root of T.

is, only if edge $x \longrightarrow u$ followed by a shortest path from u to $sink(G)$ yields a path shorter than the shortest path currently known from x to $sink(G)$. In other words, a vertex x is added to PriorityQueue only if it is an affected vertex. In effect, the algorithm avoids constructing the unshaded part of the tree T in Figure 4.4.

During this process, the set of all affected vertices is identified and every affected vertex is assigned its correct value finally. If v is affected, it is assigned its correct value in line [5]; any other affected vertex x will be assigned its correct value in line [18]. Simultaneously, the algorithm also updates the set of edges $SP(G)$ as follows. If v is unaffected but $v \longrightarrow w$ becomes an SP edge, it is added to $SP(G)$ in line [8]. Similarly any edge $x \longrightarrow u$ that becomes an SP edge, while x is unaffected, is identified and added to $SP(G)$ in line [21]. For any affected vertex u, an edge $u \longrightarrow x$ directed away from u can change its SP edge status. These changes are identified and made to $SP(G)$ in lines [12]-[15].

Note that unlike procedure DeleteEdge$_{SSSP>0}$, in which the process of identifying which vertices are members of AFFECTED and the process of updating *dist* values are separated into separate phases, in procedure InsertEdge$_{SSSP>0}$ the identification of AFFECTED is *interleaved* with updating. Observe, too, that the algorithm works correctly even if the length of the newly inserted edge is negative, as long as all other edges have a positive length and the new edge does not introduce a cycle of negative length. The reason is that we require edges to have a non-negative length only in the (partial) construction of the tree T. But in constructing a shortest-path tree for some sink vertex, one can always ignore edges going *out of the sink vertex*, as long as there are no negative length cycles. Consequently, it is immaterial, in the construction of T, whether $length(v \longrightarrow w)$ is negative or not.

We now analyze the time complexity of InsertEdge$_{SSSP>0}$. The loop in lines [10]-[24] iterates once for every affected vertex u. Each iteration takes time $O(log \, |\text{AFFECTED}|)$ for line [11] and time $O(\| \{u\} \|)$ for lines [12]-[23]. Note that the AdjustHeap operation in line [19] either inserts a vertex into the heap or decreases the key of a vertex in the heap. Hence it costs only $O(1)$ time. Thus, the running time of procedure InsertEdge$_{SSSP>0}$ is $O(\| \text{AFFECTED} \| + |\text{AFFECTED}| \, log \, |\text{AFFECTED}|)$, which is $O(\| \delta \| \, log \, \| \delta \|)$.

4.1.3. Handling Edges of Non-Positive Length

It is well known that Dijkstra's shortest-path algorithm cannot be used for graphs with negative-length edges. The presence of zero-length edges, however, poses no problem for Dijkstra's algorithm. In contrast, the obstacle to updating the SSSP solution in a bounded fashion is not the presence of negative-length edges *per se* but the presence of zero-length (or negative-length) *cycles*.

Procedure InsertEdge$_{SSSP>0}$ will work correctly even in the presence of zero-length edges, as can be verified easily. Procedure DeleteEdge$_{SSSP>0}$, however, may not work correctly if the input graph has zero-length cycles because the SP graph need no longer be a dag. (Every edge in a zero-length cycle will be in the SP graph, provided the sink is reachable from the cycle.) If the SP graph is not a dag, then

procedure InsertEdge$_{SSSP>0}$(G, $v \rightarrow w$, c)
declare
      G: a directed graph
      $v \rightarrow w$: an edge to be inserted in G
      c: a positive real number indicating the length of edge $v \rightarrow w$
      PriorityQueue: a heap of vertices
preconditions
      $SP(G)$ is the shortest-paths subgraph of G
      $\forall v \in V(G)$, $outdegree_{SP}(v)$ is the outdegree of vertex v in $SP(G)$
      $\forall v \in V(G)$, $dist(v)$ is the length of the shortest path from v to $sink(G)$
begin
[1]   Insert edge $v \rightarrow w$ into $E(G)$
[2] $length(v \rightarrow w) := c$
[3]   PriorityQueue $:= \varnothing$
[4]   **if** $length(v \rightarrow w) + dist(w) < dist(v)$ **then**
[5] $dist(v) := length(v \rightarrow w) + dist(w)$
[6]   InsertHeap(PriorityQueue, v, 0)
[7]   **else if** $length(v \rightarrow w) + dist(w) = dist(v)$ **then**
[8]   Insert $v \rightarrow w$ into $SP(G)$ and increment $outdegree_{SP}(v)$
[9]   **fi**
[10]   **while** PriorityQueue $\neq \varnothing$ **do**
[11] $u :=$ FindAndDeleteMin(PriorityQueue)
[12]   Remove all edges of $SP(G)$ directed away from u and set $outdegree_{SP}(u) = 0$
[13]   **for** every vertex $x \in Succ(u)$ **do**
[14]   **if** $length(u \rightarrow x) + dist(x) = dist(u)$ **then**
           Insert $u \rightarrow x$ into $SP(G)$ and increment $outdegree_{SP}(u)$ **fi**
[15]   **od**
[16]   **for** every vertex $x \in Pred(u)$ **do**
[17]   **if** $length(x \rightarrow u) + dist(u) < dist(x)$ **then**
[18] $dist(x) := length(x \rightarrow u) + dist(u)$
[19]   AdjustHeap(PriorityQueue, x, $dist(x) - dist(v)$)
[20]   **else if** $length(x \rightarrow u) + dist(u) = dist(x)$ **then**
[21]   Insert $x \rightarrow u$ into $SP(G)$ and increment $outdegree_{SP}(x)$
[22]   **fi**
[23]   **od**
[24]   **od**
end
postconditions
      $SP(G)$ is the shortest-paths subgraph of G
      $\forall v \in V(G)$, $outdegree_{SP}(v)$ is the outdegree of vertex v in $SP(G)$
      $\forall v \in V(G)$, $dist(v)$ is the length of the shortest path from v to $sink(G)$

Figure 4.5. An algorithm to update the SSSP>0 solution and $SP(G)$ after the insertion of an edge $v \rightarrow w$ into graph G.

Phase 1 of DeleteEdge$_{SSSP>0}$ may not identify the set of affected vertices correctly. When edge $v \rightarrow w$ is deleted from the graph shown in Figure 4.6, for example, vertices v and u are affected. But Phase 1 is unable to determine this because vertex $v \rightarrow w$ is not the only SP edge going out of v. The problem lies essentially in handling cycles in the dynamic reachability problem. We show in Chapter 8 that there exists *no* bounded incremental algorithm, within a certain model of computation, for maintaining shortest paths if zero-length cycles are allowed in the graph.

However, bounded incremental algorithms *do* exist for the dynamic SSSP-Cycle>0 problem: the dynamic single-sink shortest-path problem in graphs where edges may have arbitrary length but all cycles have positive length. First, it can be verified easily that both DeleteEdge$_{SSSP>0}$ and InsertEdge$_{SSSP>0}$ work correctly even in the presence of zero-length *edges* as long as there are no zero-length *cycles*. However, they do not work correctly in the presence of negative-length edges for the same reasons that Dijkstra's algorithm does not. We will now see how these incremental algorithms for the SSSP>0 problem can be adapted to work for the SSSP-Cycle>0 problem, often with no increase in the time complexity.

The idea is to use the technique of Edmonds and Karp for transforming the length of every edge in a graph to a non-negative real without changing the graph's shortest paths [Edm72, Tar83]. Their technique is based on the observation that if the length of each edge $a \rightarrow b$ is replaced by $f(b) + length(a \rightarrow b) - f(a)$, where f is any function that maps vertices of the graph to reals, then the shortest paths in the graph are unchanged from the original edge-length mapping (though their lengths will change). The reason is that if P is any path from a vertex a to a vertex b, then $length'(P) = f(b) + length(P) - f(a)$, where $length'$ denotes the lengths under the modified edge-length mapping. Consequently, $dist'(a,b) = f(b) + dist(a,b) - f(a)$, where $dist$ denotes the length of the shortest path between two vertices.

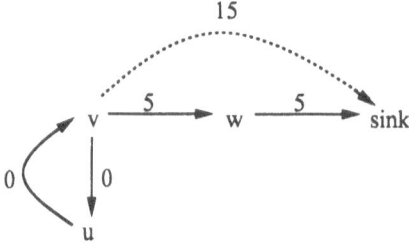

Figure 4.6. An example to show the difficulty in updating the SSSP solution after the deletion of an edge, if zero-length cycles are present in the graph. *SP* edges are represented by solid edges, and other edges by dashed edges. It appears as though vertex v is not affected after the deletion of edge $v \rightarrow w$ because of an alternative shortest path via edge $v \rightarrow u$.

Hence, if we can find some function f that satisfies

$$f(b) + length(a \rightarrow b) - f(a) \geq 0 \qquad (\dagger)$$

for every edge $a \rightarrow b$ in the graph, then Dijkstra's algorithm can be used to compute the shortest paths under the modified edge-length mapping, and the lengths of the shortest paths under the original edge-length mapping can be easily recovered from this information. Our goal is to use this technique for the *dynamic* SSSP problem.

We note that such a function f is available for the problem of incrementally updating shortest paths to a sink after the deletion of an edge: we simply let $f(u)$ be $dist_{old}(u)$, the length of the shortest path in graph G from u to $sink(G)$ before G was modified. Note that for every edge $a \rightarrow b$ in the original graph $dist_{old}(a) \leq dist_{old}(b) + length(a \rightarrow b)$, and hence, $dist_{old}$ meets the properties required of f. This is not completely true since $dist_{old}(u)$ is ∞ for any vertex u from which the sink cannot be reached. However, the deletion of an edge $v \rightarrow w$, where $dist_{old}(v)$ is ∞, causes no change in the solution, and is trivially handled. If, on the other hand, $dist_{old}(v)$ is finite, then $dist_{old}(u)$ is finite for every vertex u that is visited during the update. Vertices that could not reach the sink in the original graph are of no interest. Hence, $dist_{old}$ turns out to be a satisfactory choice for the function f. The update algorithm works just like DeleteEdge$_{SSSP>0}$, except that it uses the rescaled functions *length'* and *dist'* instead of *length* and *dist*. The new *dist* values can be easily computed from the updated *dist'* values as indicated above.

We use the same strategy for updating the solution after the insertion of an edge $v \rightarrow w$. However, in the case of an edge insertion, the function $dist_{old}$ can fail to satisfy property ($\dagger$) in two cases. The first case is that the transformed length of the newly inserted edge $v \rightarrow w$ is not guaranteed to be non-negative; however, this causes no difficulties because, as explained in the previous section, InsertEdge$_{SSSP>0}$ works correctly even if the length of the inserted edge is non-positive.

The second case is more problematic. The above technique for updating the solution to the SSSP-Cycle>0 problem fails for one type of input change, namely the insertion of an edge $v \rightarrow w$ that creates a path from v to the sink vertex where no path existed before. For such an input modification, we cannot use $dist_{old}$ as the function f, since $dist_{old}$ is ∞ for vertex v and other vertices from which the sink was unreachable in the original graph. However, even such an input modification can be handled in time $O(\|\delta\| \cdot |\delta|)$ as follows. Note that in what follows we assume that $dist_{old}(v)$ is ∞ while $dist_{old}(w)$ is finite.

Consider the set U of all vertices that reach v for which $dist_{old}$ is ∞. Solve the single-sink shortest-path problem for the subgraph induced by U with v as the sink using the Bellman-Ford algorithm or any other suitable shortest-path algorithm. (See [Cor90], for example.) This can be done in $O(\|\delta\| \cdot |\delta|)$ time since every vertex in U is an affected vertex. This gives us the length $dist(u,v)$ of the shortest path from u to v for every vertex u in U. Now, define the function f by:

$$f(u) = dist_{old}(u) \qquad \text{if } u \notin U$$
$$ = dist(u,v) + C \qquad \text{otherwise}$$

where C is a constant chosen so that the rescaled lengths of all edges except $v \rightarrow w$

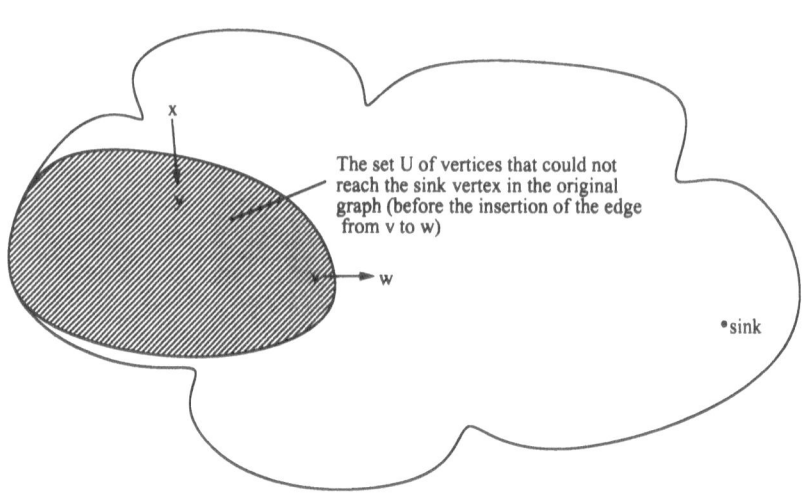

Figure 4.7. Updating the shortest path information, in the presence of negative-length edges, after the insertion of an edge $v \rightarrow w$, when $dist_{old}(v)$ is ∞ and $dist_{old}(w)$ is finite. The shaded region in the above figure is the set U of vertices that can reach vertex v but not the sink vertex in the original graph.

are non-negative. Thus, C should be chosen such that for all edges $x \rightarrow y$ with x out-side U and y in U, $f(y) + length(x \rightarrow y) - f(x) \geq 0$. Substituting from the definition of f, C should be such that $dist(y,v) + C + length(x \rightarrow y) - dist_{old}(x) \geq 0$. Hence, we define C to be max $\{ dist_{old}(x) - length(x \rightarrow y) - dist(y,v) \mid x \rightarrow y \in E(G)$ and $x \notin U$ and $y \in U \}$. Now, we can use an adaptation of InsertEdge$_{SSSP>0}$ that makes use of the rescaled lengths to perform the update to the solution.

We can think of the above method as inserting the edge $v \rightarrow w$ with its correct length in two steps: first, the edge $v \rightarrow w$ is inserted with a length large enough that only vertices in U are affected; the updating for this input modification is performed using an adaptation of the Bellman-Ford algorithm; then, the length of the edge $v \rightarrow w$ is reduced to its correct length; the updating for this modification is then done using the previously explained methods.

4.1.4. A Batch SSSP Algorithm

We have so far restricted our attention to unit changes. We will address the problem of updating the solution to an instance of the SSSP>0 problem after an arbitrary change in the input graph in Chapter 6. We will meanwhile consider a special class of non-unit changes that can be handled easily using the incremental algorithms that have been presented in this chapter and discuss an application of these algorithms.

Consider the insertion of a number of edges $v \rightarrow w_1, \cdots, v \rightarrow w_i$, all directed away from the same vertex v, into the graph. Each edge $v \rightarrow w_j$ introduces a candidate for the new shortest path from v to the sink, whose length is given by $length(v \rightarrow w_j) + dist(w_j)$. We can easily modify InsertEdge$_{\text{SSSP}>0}$ to handle an input change of this kind: we would insert all these edges into the graph, compare all candidate shortest paths to identify the best candidate, and then proceed as before.

We now consider an application of this generalized incremental algorithm to the batch SSSP problem. Yap [Yap83] describes an algorithm for finding the shortest path between two vertices in a graph that may include edges with negative length. This algorithm works better than the standard Bellman-Ford algorithm when the number of negative-length edges is small. We now present an algorithm for this problem that has a better time complexity. This algorithm, in fact, solves the single-source or single-sink problem rather than just the single-pair problem.

We first consider the time complexity of Yap's algorithm. Let G be the given graph. Let n denote the number of vertices in G and let m denote the number of edges in G. Let h denote the number of edges whose length is negative, and let k denote $\min(h,n)$. Yap's approach reduces a single-pair shortest path problem on the given graph G to $\min(h+1,n)$ SSSP≥ 0 problems on the subgraph of G consisting of only non-negative edges, and a single-pair shortest-path problem on a graph consisting of $O(k)$ vertices and $O(k^2)$ edges of arbitrary (that is, both positive and negative) lengths. This yields an $O(k[m + n \log n] + k^3)$ algorithm for the problem, which is better than the standard $O(mn)$ algorithm for sufficiently small k. (Actually, Yap describes the time complexity of the algorithm as $O(kn^2)$, since he makes use of Dijkstra's $O(n^2)$ algorithm. The above complexity follows from Fredman and Tarjan's [Fre87] improvement to Dijkstra's algorithm. The complexity of the above algorithm can be improved slightly by utilising the recent $O(m + n \log n / \log\log n)$ shortest path algorithm due to Fredman and Willard [Fre90a]).

We now consider how our incremental algorithm for the shortest-path problem can be used to solve this problem better. Let $u_1, \ldots, u_{k'}$ be the set of all vertices in the graph that have an outgoing edge of negative length. Thus $k' \leq k$. First replace all the negative edges in the given graph G with zero-length edges. Compute the solution to this graph by using, say, the Fredman-Tarjan improvement to Dijkstra's algorithm. Now process the vertices $u_1, \ldots, u_{k'}$ one by one. The vertex u_i is processed by restoring the lengths of all the edges directed away from u_i to their actual values and updating the solution using the generalization of InsertEdge$_{\text{SSSP}>0}$ discussed above.

The updating after each insertion step takes $O(\|\delta\| + |\delta| \log |\delta|)$ time, which is $O(m + n \log n)$ time in the worst case. Hence, the algorithm runs in time $O(k'[m + n \log n])$. In general, the algorithm can be expected to take less time than this bound, since all the update steps have bounded complexity.

4.2. The Dynamic All-Pairs Shortest-Path Problem

This section concerns a bounded incremental algorithm for a version of the dynamic all-pairs shortest-path problem with positive-length edges (APSP>0).

We will assume that the vertices of G are indexed from $1 .. |V(G)|$. APSP>0 involves computing the entries of a *distance matrix*, $dist[1 .. |V(G)|, 1 .. |V(G)|]$, where entry $dist[i, j]$ represents the length of the shortest path in G from vertex i to vertex j. It is also useful to think of this information as being associated with the individual vertices of the graph: with each vertex there is an *array* of values, indexed from $1 .. |V(G)|$—the jth value at vertex i records the length of the shortest path in G from vertex i to vertex j. This lets us view the APSP>0 problem as a graph problem that requires the computation of an output value for each vertex in the graph, though the actual information may be stored in a two-dimensional array, if necessary. However, APSP>0 does not fall into the class of graph problems that involve the computation of a *single atomic* value for each vertex u in the input graph, and so, as explained below, some of our terminology in this section differs from the terminology that was introduced in Chapter 3.

Since MODIFIED measures the change in the input, the definition of MODIFIED remains the same (and hence for a single-edge change to the graph $|$MODIFIED$| = 2$). In order to define AFFECTED, which measures the change in the output, we view the problem as n instances of the SSSP>0 problem. Let AFFECTED$_u$ represent the set of affected vertices for the single-sink problem with u as the sink vertex. We define $|$AFFECTED$|$ for the APSP>0 problem as follows:

$$|\text{AFFECTED}| = \sum_{u = 1}^{|V(G)|} |\text{AFFECTED}_u|.$$

Thus, $|$AFFECTED$|$ is the number of entries in the *dist* matrix that change in value. We define the extended size $\|$AFFECTED$\|$ as follows:

$$\|\text{AFFECTED}\|_i = \sum_{u = 1}^{|V(G)|} \|\text{AFFECTED}_u\|_i,$$

Note that for a given change δ, some or all of the AFFECTED$_u$ can be empty and, hence, $\|$AFFECTED$\|_i$ may be less than $|V(G)|$. The parameter $\|\delta\|_i$ in which we measure the incremental complexity of APSP>0 is defined as follows:

$$\|\delta\|_i = \|\text{MODIFIED}\|_i + \|\text{AFFECTED}\|_i.$$

The parameter $|\delta|$ is also similarly defined.

Though it is arguable how precisely $\|\delta\|$, as defined above, measures the size of the change in the output, the above definitions are clearly in the same spirit as those from Chapter 3.

We now turn our attention to the problem of updating the solution to an instance of the APSP>0 problem after a unit change.

The operations of inserting and deleting isolated vertices are trivially handled but for some concerns having to do with dynamic storage allocation. Whether the shortest-path distances are stored in a single two-dimensional array or in a collection of one-dimensional arrays, we face the need to *increase* or *decrease* the array size(s). We can do this by dynamically expanding and contracting these arrays using the well-known doubling/halving technique (see Section 18.4 of [Cor90], for example). Assume the distance matrix is maintained as a collection of n vectors (of equal size), where n is the number of vertices in the graph. Whenever a new vertex is inserted, a new vector is allocated. Whenever the number of vertices in the graph exceeds the size of the individual vectors, the size of each of the vectors is doubled (by re-allocation). Vertex deletion is similarly handled, by halving the size of the vectors when appropriate. The insertion or deletion of an isolated vertex has an *amortized* cost of $O(|V(G)|)$ under this scheme: doubling or halving the arrays takes time $O(|V(G)|^2)$, but the cost is amortized over $\Omega(|V(G)|)$ vertex insertion/deletion operations. A cost of $O(|V(G)|)$ is reasonable, in the sense that the introduction or removal of an isolated vertex causes $O(|V(G)|)$ "changes" to entries in the distance matrix. Thus, in some sense for such operations $|\delta| = \Theta(|V(G)|)$, and hence the amortized cost of the doubling/halving scheme is optimal.

We now consider the problem of updating the solution after the insertion or deletion of an edge. As explained in the previous section, it is trivial to generalize these operations to handle the shortening or lengthening of an edge, respectively.

Proposition 4.2. *APSP>0 has a bounded incremental algorithm. In particular, there exists an algorithm* DeleteEdge$_{APSP>0}$ *that can process an edge deletion in time* $O(\|\delta\|_2 + |\delta| \log |\delta|)$, *and there exists an algorithm* InsertEdge$_{APSP>0}$ *that can process an edge insertion in time* $O(\|\delta\|_1)$.

4.2.1. Deletion of an Edge

We can view an instance of the APSP>0 problem as n instances of the SSSP>0 problem. The basic idea behind the bounded incremental algorithm for DeleteEdge$_{APSP>0}$ is to make repeated use of the bounded incremental algorithm DeleteEdge$_{SSSP>0}$ as a subroutine, but with a different sink vertex on each call. A simple incremental algorithm for DeleteEdge$_{APSP>0}$ would be to make as many calls on DeleteEdge$_{SSSP>0}$ as there are vertices in graph G. However, this method is not bounded because it would perform at least *some* work for each vertex of G; the total updating cost would be at least $O(|V(G)|)$, which in general is not a function of $\|\delta\|_i$ for any fixed value of i.

The key observation behind our bounded incremental algorithm for DeleteEdge$_{APSP>0}$ is that it is possible to determine *exactly* which calls on DeleteEdge$_{SSSP>0}$ are necessary. With this information in hand it is possible to keep the total updating cost bounded. Let us call a vertex y an *affected sink* if there exists some vertex x such that $dist(x,y)$ changes as a consequence of the edge deletion. Our algorithm works by first computing the set of affected sink vertices, and then invoking DeleteEdge$_{SSSP>0}$ for each affected sink vertex.

If $dist(x,y)$ changes as a consequence of the deletion of an edge $v \longrightarrow w$, then all shortest paths from x to y in the original graph must have passed through edge $v \longrightarrow w$. This implies that all shortest paths from v to y must have passed through edge $v \longrightarrow w$ as well. Hence, $dist(v,y)$ must also change as a consequence of the edge deletion. In other words, y is an affected sink following the deletion of an edge $v \longrightarrow w$ iff $dist(v,y)$ changes as a result of the edge-deletion. Hence, the set of affected sink vertices is the same as the set of affected vertices for the *single-source* shortest-path problem with v as the source vertex. We know how to compute this set from our edge-deletion algorithm for the SSSP>0 problem. In particular, let G^R

denote the graph obtained from G by reversing all edges in G. The set of affected vertices identified in Phase 1 when we run the SSSP>0 edge-deletion algorithm for sink v in the graph G^R yields AffectedSinks. This is what we do in line [2] of procedure DeleteEdge$_{APSP>0}$. (See Figure 4.9.) Obviously there is no need to explicitly construct the graph G^R—we use it in the algorithm description merely to simplify the presentation.

Once we have identified the affected sinks, we just need to invoke DeleteEdge$_{SSSP>0}$ for each affected sink. However, we do not use procedure DeleteEdge$_{SSSP>0}$ itself, but an adaptation of that procedure. This adaptation, presented as procedure DeleteUpdate in Figure 4.8, differs from DeleteEdge$_{SSSP>0}$ in the following ways. First, it accepts an extra argument, vertex z, which is the sink vertex for which the SSSP>0 solution needs to be updated. Second, unlike DeleteEdge$_{SSSP>0}$, DeleteUpdate does not delete the edge from the graph, since the deletion of edge $v \rightarrow w$ is performed in DeleteEdge$_{APSP>0}$ itself (see line [1]). Instead, it updates only the shortest-path distances to sink z. Third, and most importantly, the algorithm does not maintain either an explicit representation of the SP dag or the outdegree of vertices in the SP dag, for the reasons described below.

Note that there is one *SP* dag for each vertex in the graph. Consequently, the amount of *SP* information that changes (for the entire collection of different sinks) can be unbounded for certain edge-modification operations. For example, when an edge $v \rightarrow w$ is inserted with a length equal to $dist(v, w)$, none of the entries in the distance matrix change value, and consequently

$$\| \delta \|_2 = \| \text{MODIFIED} \|_2 + \sum_{u=1}^{|V(G)|} \| \text{AFFECTED}_u \|_2$$

$$= \| \text{MODIFIED} \|_2.$$

Such an insertion can introduce a new element in the *SP* set for each of the different sinks, and thus cause a change in *SP* information of size $\Omega(|V(G)|)$. Thus, maintaining the SP graphs for every sink would be too expensive, and would result in an unbounded incremental algorithm. The algorithm uses, instead, the predicate $SP(a, b, c)$:

$$SP(a, b, c) \equiv (dist(a, c) = length(a \rightarrow b) + dist(b, c)) \wedge (dist(a, c) \neq \infty).$$

Predicate $SP(a, b, c)$ answers the question "Is edge $a \rightarrow b$ an *SP* edge when vertex c is the sink?" This check can be done in constant time.

The use of predicate $SP(a, b, c)$ makes it important that the test in line [10] be carefully implemented. Recall that Phase 1 is similar to a (reverse) topological order traversal in the SP dag for sink z. We are interested in determining in line [10] if every successor of x in the SP dag has already been "visited" and placed in AffectedVertices; if so, then x can be placed in AffectedVertices too. In procedure DeleteEdge$_{SSSP>0}$ we used the standard technique for performing a topological order traversal: a count was maintained at each vertex of the number of its successors (in the SP dag) not yet placed in AffectedVertices; when the count for a vertex x fell to zero, it was placed in the WorkSet.

---

function DeleteUpdate(G, $v \rightarrow w$, z) **returns** a set of vertices
declare
      G: a directed graph
      $v \rightarrow w$: the edge that has been deleted from G
      z: the sink vertex of G
      WorkSet, AffectedVertices: sets of vertices
      PriorityQueue: a heap of vertices
      $SP(a, b, c) \equiv (dist_G(a, c) = length_G(a \rightarrow b) + dist_G(b, c)) \wedge (dist_G(a, c) \neq \infty)$
begin
[1]   AffectedVertices := $\emptyset$
[2]   **if** there does not exist any vertex $x \in Succ_G(v)$ such that $SP(v, x, z)$ **then**
[3]   /* Phase 1: Identify affected vertices */
[4]   WorkSet := { v }
[5]   **while** WorkSet $\neq \emptyset$ **do**
[6]   Select and remove a vertex u from WorkSet
[7]   Insert vertex u into AffectedVertices
[8]   **for** each vertex $x \in Pred_G(u)$ such that $SP(x, u, z)$ **do**
[9]   **if** $\nexists y \in Succ_G(x) -$ AffectedVertices such that $SP(x, y, z)$
                 then Insert vertex x into WorkSet **fi**
[10]   **od**
[11]   **od**
[12]   /* Phase 2: Determine new distances to z for all vertices in AffectedVertices. */
[13]   PriorityQueue := $\emptyset$
[14]   **for** each vertex $a \in$ AffectedVertices **do**
[15] $dist_G(a, z) := \min (\{ length_G(a \rightarrow b) + dist_G(b, z) \mid$
                        $a \rightarrow b \in E(G)$ and $b \notin$ AffectedVertices) $\} \cup \{ \infty \})$
[16]   **if** $dist_G(a, z) \neq \infty$ **then** InsertHeap(PriorityQueue, a, $dist_G(a, z)$) **fi**
[17]   **od**
[18]   **while** PriorityQueue $\neq \emptyset$ **do**
[19] $a :=$ FindAndDeleteMin(PriorityQueue)
[20]   **for** every $c \in Pred_G(a)$ such that $length_G(c \rightarrow a) + dist_G(a, z) < dist_G(c, z)$ **do**
[21] $dist_G(c, z) := length_G(c \rightarrow a) + dist_G(a, z)$
[22]   AdjustHeap(PriorityQueue, c, $dist_G(c, z)$)
[23]   **od**
[24]   **od**
[25]   **fi**
[26]   return AffectedVertices
end

Figure 4.8. Procedure DeleteUpdate updates distances to vertex z after edge $v \rightarrow w$ is deleted from G.

Since we cannot afford to maintain a similar count (across updates to the graph), we need to perform the check in line [10] differently. Note that the check in line [10] can be performed multiple times for the *same* vertex x. In fact, a vertex x can be checked *outdegree*(x) times. If we examine all successors of vertex x each time, the cost of the repeated checks in line [10] for a *particular* vertex x can be quadratic in the number of successors it has. Instead, the same total cost can be made linear in *outdegree* (x) by using one of the following strategies.

The first time vertex x is checked in line [10] we count the number of vertices y in $Succ(x)$ – AffectedVertices that satisfy $SP(x,y,z)$. Whenever vertex x is subsequently checked in line [10] we decrement its count. We add x to the WorkSet when its count falls to zero.

Alternatively, we can represent the set of successors of a vertex as a linear linked list. The test in line [10] requires us to scan this list to find a vertex y satisfying $SP(x,y,z)$ that is yet not in AffectedVertices. If we store a pointer, in vertex x, to the portion of this list that has not been scanned yet, then during subsequent tests on vertex x we can resume the scan where we left it.

Even these tricks do not make the algorithm bounded in $\|\delta\|_1$. The reason is that the vertex x checked in line [10] is not necessarily a member of AFFECTED, but we are forced to examine all successors of x. However, even if the tested vertex x is not a member of AFFECTED it is guaranteed to be a predecessor of a member of AFFECTED. Consequently, the algorithm is bounded in $\|\delta\|_2$. In particular, the cost of Phase 1 is bounded by $O(\|\text{MODIFIED}\|_1 + \|\text{AFFECTED}_z\|_2)$; The cost of Phase 2 is bounded by $O(\|\text{AFFECTED}_z\|_1 + |\text{AFFECTED}_z| \log |\text{AFFECTED}_z|)$.

Consequently, the cost of DeleteEdge$_{\text{APSP}>0}$ is $O(\|\delta\|_2 + |\delta| \log |\delta|)$.

4.2.2. Insertion of an Edge

We now present a bounded incremental algorithm for the problem of updating the solution to APSP>0 after an edge $v \rightarrow w$ of length c is inserted into G. Though similar bounded algorithms have been previously proposed for this problem (see Rohnert

---

procedure DeleteEdge$_{\text{APSP}>0}(G, v \rightarrow w)$
declare
      G: a directed graph
      $v \rightarrow w$: an edge to be deleted from G
      AffectedSinks: a set of vertices
preconditions
      $\forall u, v \in V(G), dist(u,v)$ is the length of the shortest path from u to v
begin
[1]   Remove edge $v \rightarrow w$ from $E(G)$
[2]   AffectedSinks := DeleteUpdate(G^R, $w \rightarrow v$, v)
[3]   **for** each vertex $x \in$ AffectedSinks **do** DeleteUpdate(G, $v \rightarrow w$, x) **od**
end
postconditions
      $\forall u, v \in V(G), dist(u,v)$ is the length of the shortest path from u to v

---

Figure 4.9. Procedure DeleteEdge$_{\text{APSP}>0}$ updates the solution to APSP>0 after edge $v \rightarrow w$ is deleted from G.

[Roh85], Even and Gazit [Eve85], Lin and Chang [Lin90], and Ausiello *et al.* [Aus90]), we present the algorithm for the sake of completeness. We observe that the algorithms described by Rohnert [Roh85], Lin and Chang [Lin90], and Ausiello *et al.* [Aus90]) all maintain a shortest-path-tree data structure for each vertex, which makes their edge-insertion algorithm somewhat faster. However, maintaining the shortest-path-tree makes the processing of an *edge deletion* more expensive (and unbounded).

   As in the case of edge deletion, we may obtain a bounded incremental algorithm for edge insertion as follows: compute AffectedSinks, the set of all vertices y for which there exists a vertex x such that the length of the shortest path from x to y has changed; for every vertex y in AffectedSinks, invoke the bounded incremental operation InsertEdge$_{SSSP>0}$ with y as the sink.

   The algorithm InsertEdge$_{APSP>0}$ presented in Figure 4.11 carries out essentially the technique outlined above, but with one difference. It makes use of a considerably simplified form of the procedure InsertEdge$_{SSSP>0}$, which is given as procedure

---

function InsertUpdate(G, $v \rightarrow w$, z) **returns** a set of vertices
declare
       G: a directed graph
       $v \rightarrow w$: the edge that has been inserted in G
       z: the sink vertex of G
       WorkSet: a set of edges
       VisitedVertices: a set of vertices
       $SP(a, b, c) \equiv (dist_G(a, c) = length_G(a \rightarrow b) + dist_G(b, c)) \wedge (dist_G(a, c) \neq \infty)$
begin
[1]   WorkSet := { $v \rightarrow w$ }
[2]   VisitedVertices := { v }
[3]   AffectedVertices := $\varnothing$
[4]   **while** WorkSet $\neq \varnothing$ **do**
[5]   Select and remove an edge $x \rightarrow u$ from WorkSet
[6]   **if** $length_G(x \rightarrow u) + dist_G(u,z) < dist_G(x,z)$ **then**
[7]   Insert x into AffectedVertices
[8] $dist_G(x,z) := length_G(x \rightarrow u) + dist_G(u,z)$
[9]   **for** every vertex $y \in Pred_G(x)$ **do**
[10]   **if** $SP(y,x,v)$ **and** $y \notin$ VisitedVertices **then**
[11]   Insert $y \rightarrow x$ into WorkSet
[12]   Insert y into VisitedVertices
[13]   **fi**
[14]   **od**
[15]   **fi**
[16]   **od**
[17]   return AffectedVertices
end

Figure 4.10. Procedure InsertUpdate updates distances to vertex z after edge $v \rightarrow w$ is inserted into G.

InsertUpdate in Figure 4.10. The simplifications incorporated in InsertUpdate are explained below.

Recall the description of InsertEdge$_{SSSP>0}$ given in Section 3.1.2. InsertEdge$_{SSSP>0}$ makes use of an adaptation of Dijkstra's algorithm to identify shortest paths to sink v and update distance information. However, in InsertUpdate, the dag of all shortest paths to sink v is already available (albeit in an implicit form), and this information can be exploited to sidestep the use of a priority queue. (Note that the insertion of the edge $v \rightarrow w$ cannot affect shortest paths to sink v, since the graph contains no cycles of negative length. Hence, the dag of shortest paths to sink v undergoes no change during InsertEdge$_{APSP>0}$.) As explained in Section 3.2.1, the predicate $SP(a,b,v)$ can be used to determine, in constant time, if the edge $a \rightarrow b$ is part of the dag of shortest paths to sink v. This permits InsertUpdate to do a (partial) backward traversal of this dag, visiting only affected vertices or their predecessors.

For instance, consider the edge $x \rightarrow u$ selected in line [5] of Figure 4.10. Vertex x is the vertex to be visited next during the traversal described above. Except in the case when edge $x \rightarrow u$ is $v \rightarrow w$, vertex u is an affected vertex and is the successor of x in a shortest path from x to v. The test in line [6] determines if x itself is an affected vertex. If it is, its distance information is updated, and its predecessors in the shortest-path dag to sink v are added to the workset for subsequent processing, unless they have already been visited. The purpose of the set VisitedVertices is to keep track of all the vertices visited in order to avoid visiting any vertex more than once. For reasons to be given shortly, InsertUpdate simultaneously computes AffectedVertices, the set of all vertices the length of whose shortest path to vertex z changes.

---

procedure InsertEdge$_{APSP>0}(G, v \rightarrow w, c)$
declare
        G: a directed graph
        $v \rightarrow w$: an edge to be inserted in G
        c: a positive real number indicating the length of edge $v \rightarrow w$
        AffectedSinks: a set of vertices
preconditions
        $\forall u, v \in V(G)$, $dist(u,v)$ is length of the shortest path from u to v
begin
[1]   Insert edge $v \rightarrow w$ into $E(G)$
[2] $length_G(v \rightarrow w) := c$
[3]   AffectedSinks := InsertUpdate($G^R, w \rightarrow v, v$)
[4]   **for** each vertex $x \in$ AffectedSinks **do** InsertUpdate($G, v \rightarrow w, x$) **od**
end
postconditions
        $\forall u, v \in V(G)$, $dist(u,v)$ is length of the shortest path from u to v

---

Figure 4.11. Procedure InsertEdge$_{APSP>0}$ updates the solution to APSP>0 after edge $v \rightarrow w$ of length c is inserted in G.

We now justify the method used in InsertEdge$_{APSP>0}$ to determine Affected-Sinks, the set of all vertices y for which there exists a vertex x such that the length of the shortest path from x to y has changed. This set is the set of sinks for which InsertEdge$_{APSP>0}$ must invoke InsertUpdate. Assume that x and y are vertices such that the length of the shortest path from x to y changes following the insertion of edge $v \longrightarrow w$. Then, the new shortest path from x to y must pass through the edge $v \longrightarrow w$. Obviously, the length of the shortest path from v to y must have changed as well. Hence, AffectedSinks is the set { y | the length of the shortest path from v to y changes following the insertion of edge $v \longrightarrow w$ }. This set is precisely the set of all affected vertices for the single-source shortest-path problem with v as the source, *i.e.*, the set AffectedVertices computed by the call InsertUpdate($G^R, w \longrightarrow v, v$). This is how InsertEdge$_{APSP>0}$ determines the set AffectedSinks (see line [3] of Figure 4.11); InsertUpdate is then invoked repeatedly, once for each member of AffectedSinks.

We now consider the time complexity of InsertEdge$_{APSP>0}$. Note that for every vertex $x \in$ AffectedSinks, any vertex examined by InsertUpdate($G, v \longrightarrow w, x$) is in $N(\text{AFFECTED}_x)$. InsertUpdate does essentially a simple traversal of the graph $<N(\text{AFFECTED}_x)>$, in time $O(\| \text{AFFECTED}_x \|)$. Thus, the total running time of line [4] in procedure InsertEdge$_{APSP>0}$ is $O(\| \delta \|_1)$. Thus, the total running time of procedure InsertEdge$_{APSP>0}$ is $O(\| \delta \|_1)$.

4.2.3. Handling Edges of Non-Positive Length

The techniques described earlier, in Section 4.1.3, for handling negative edge lengths in the case of DeleteEdge$_{SSSP}$ carry over to DeleteEdge$_{APSP}$ too. Note that InsertEdge$_{APSP>0}$, as presented, works correctly even in the presence of negative edge-lengths. In InsertEdge$_{SSSP}$ we needed to identify shortest paths to vertex v when an edge $v \longrightarrow w$ was inserted, which we did by using an adaptation of Dijkstra's algorithm. Consequently, we had to take special care of negative-length edges. In InsertEdge$_{APSP}$, however, the shortest paths information is already available, as explained earlier. Hence, negative-length edges are no problem in the case of InsertEdge$_{APSP}$.

4.3. Related Work

Various versions of the dynamic shortest-path problem have attracted wide attention, beginning with Murchland's paper in 1967 [Mur67]. There has not been much work, however, on the dynamic single-sink or single-source shortest-path problem. Bounded incremental algorithms were previously known only for the case of an edge insertion in the all-pairs shortest-path problem [Roh85, Eve85, Lin90, Aus90, Aus91]. The incremental algorithm presented in this chapter for processing the insertion of an edge in the case of the all-pairs shortest-path problem is similar to, but was developed independently of, these bounded algorithms. A more comprehensive overview of the previous work on dynamic shortest-path problems appears in Section 6.6.

Chapter 5
Generalizations of the Shortest-Path Problem

> ··· *Survey the problem in the most natural way, taking it as solved and visualizing in suitable order all the relations that must hold between the unknowns and the data according to the conditions. Detach a part of the condition according to which you can express the same quantity in two different ways and so obtain an equation between the unknowns. Eventually you should split the condition into as many parts, and so obtain a system of as many equations, as there are unknowns.*
>
> Descartes' universal method
> —[as described in] G. Polya, *Mathematical Discovery, Volume 1*

In this chapter we look at some generalizations of the shortest-path problem. We review the *algebraic path problem* and the *dataflow analysis problem*, both of which generalize the shortest-path problem along a certain dimension, and establish the equivalence between these problems. We then review a generalization of the shortest-path problem, along a different dimension, proposed by Knuth. We then propose the *grammar* problem, a problem that combines both the above generalizations, and explore some of its applications. We study how some results relating dataflow analysis problems to maximum fixed point computation carry over to grammar problems. We then study some special grammar problems, closely related to Knuth's problem, for which we will present incremental algorithms in the next chapter.

5.1. An Overview of the Chapter

The algebraic path problem is a generalization of the shortest-path problem which has been widely studied [Car71, Zim81, Mah84, Gon84, Gon84a, Rot90]. A variety of path problems in graphs, such as those of identifying the shortest path, or the most reliable path, or the path with the maximum capacity, between certain pairs of vertices, share a common structure. An abstract version of these path problems, the algebraic path problem, can be defined in terms of the algebraic structure that is common to these problems, namely the *semiring* or *dioid*. Algorithms developed for various specific path problems have been generalized and adapted for the algebraic path problem. The advantage of studying this algebraic formulation of the problem is that algorithms developed for it can be used for any of the large number of path problems that fit into this algebraic framework.

Dataflow analysis concerns the static analysis of programs to identify properties of the analyzed programs. It has numerous applications, especially in optimizing, vectorizing or parallelizing, debugging, and testing programs. Dataflow analysis problems come in various flavors. The live-variable problem is a typical dataflow analysis problem. The input in this case consists of a control-flow graph, which represents a program. A vertex in the graph represents a basic block or a statement in the program. Each vertex u in the graph is associated with two sets of program variables: *defined*(u), the set of program variables that are assigned a value in vertex u, and *used*(u), the set of program variables that are used before being defined in vertex u. A program variable x is said to be *live* at (entry to) a vertex u iff there exists a path P from u to some vertex w such that x is in *used*(w) and x is not in *defined*(v) for any vertex v in $P-\{w\}$. The goal is to determine for every "program point" or vertex in the graph the set of all variables that are live at that point.

Kildall [Kil73] showed that the various different dataflow analysis problems have a common structure, and utilized this to define an abstract, lattice-theoretic, version of the dataflow analysis problem—the distributive dataflow analysis problem. An algorithm for this abstract problem—such as the one Kildall presented—can be utilized to solve any dataflow analysis problem that fits into the abstract framework. Some of the early work extending Kildall's work can be found in [Kam76, Gra76, Kam77]. A recent, comprehensive survey of the dataflow analysis problem appears in [Mar90a].

The title of this chapter notwithstanding, our interest in the algebraic path problem and dataflow analysis stems from the fact that these problems are generalizations of the reachability problem in graphs. We show later on, in Chapter 8, that the reachability problem is unbounded with respect to a particular model of computation. We also show there, using problem reduction, that various problems considered in this chapter are unbounded.

The first part of this chapter is primarily a review of the algebraic path problem and the dataflow analysis problem. These two frameworks are very similar—problems that can be formulated in one of these frameworks can often be reformulated in the other framework. Despite this similarity, work in these two areas has to a large extent proceeded independently, occasionally reproducing results.[1] We summarize the essential characteristics of these frameworks, and show how the two frameworks correspond to each other—that is, we show how problems formulated in one framework can be reformulated in the other framework. This section, by reviewing the several closely related algebraic systems that have been used for formulating algebraic path problems, also tries to clarify the relevance and roles of the different axioms in path algebras.

The second part of this chapter concerns the grammar problem. Both algebraic path problems and dataflow analysis problems require summarizing the set of all paths between two specific vertices in a directed graph. If the graph is viewed as a finite state machine, then the set of all paths between two specific vertices, that is, the set that is being summarized, is seen to be a regular set. This view suggests a gen-

---

[1] Tarjan, in his presentation of an algorithm for solving various path problems uniformly [Tar81, Tar81a], does discuss problems drawn from both frameworks. Gondran and Minoux [Gon84a], in their study of path algebras consider problems similar to distributive dataflow analysis under the title of generalized path algebras, but do not relate it to dataflow analysis. However, to our knowledge, the exact relationship between these two frameworks has not been formalized before.

eralization of the summary problem that requires summarizing context free languages or sets. This idea was developed in a paper by Knuth [Knu77], where he introduced a generalization of the shortest-path problem in graphs.

In this generalization, a context free grammar is given such that every derivation of a terminal string from a non-terminal is associated with a real-valued cost, and the problem is to determine for every non-terminal of the given grammar the least-cost derivation of a terminal string from that non-terminal. A simple example of this problem, which is analogous to the shortest-path problem, is that of computing for each non-terminal the shortest terminal string derivable from that non-terminal. Other examples of this problem are the nullable non-terminals problem and the useless non-terminals problem, both of which resemble the reachability problem in graphs. The nullable non-terminals problem is to identify the set of non-terminals in the input context free grammar that can derive the null string, while the useless non-terminals problem is to identify the set of non-terminals that cannot derive a terminal string. Knuth lists a variety of other applications and special cases of this problem, including the generation of optimal code for expression trees and the construction of optimal binary-search trees.

In the second part of this chapter we introduce the grammar problem, which combines ideas from the dataflow analysis problem, algebraic path problem, and Knuth's problem. In this generalization, every derivation of a terminal string from a non-terminal is associated with a value drawn from a partially ordered set, and the problem is to determine for every non-terminal of the given grammar the meet over all derivations of a terminal string from that non-terminal of the value of that derivation. The first-set computation problem—the problem of determining for each non-terminal Y the set of all terminals a such that Y can derive a terminal string beginning with a— is an example of this generalized grammar problem. Interprocedural dataflow analysis problems too can be formulated within this framework.

The problems considered in this chapter may be described as "summary problems" because they all have the following generic structure: A (possibly infinite) set or multi-set of values is specified by some means, and the problem is to compute some "summary" information of this set or multi-set of values. Most of the summary problems we consider in this paper can be reduced to the computation of the maximum fixed point, in a partially ordered set of values, of an appropriate collection of equations. We will find that the "summary" version of the problem is a natural specification of the problem, which is easy to understand, while the fixed point formulation of the problem permits us to solve the problem using standard methods for finding fixed points of equations.

The rest of this chapter is organized as follows. In Section 5.2 we review the maximum fixed point problem. In Section 5.3 we look at the algebraic path problem and dataflow analysis problem and the relationship between them. In Section 5.4 we study the grammar problem.

5.2. Maximum Fixed Point Problems

We now introduce some terminology. Let $(D, \leq)$ be a partially ordered set or poset. The poset $(D, \leq)$ is said to be *well-founded* or *bounded* if it has no infinite descending chain. The greatest element in D, if it exists, will be denoted by $\top$. The greatest lower bound (or the meet) of two elements a and b will be denoted by $a \sqcap b$. If every two elements in D have a greatest lower bound then $(D, \sqcap)$ is said to be a meet semi-lattice. The greatest lower bound of a set of elements S will be denoted by $\sqcap S$ or $\underset{s \in S}{\sqcap} s$. The semilattice is said to be complete or closed with respect to the meet operation if every subset of D has a greatest lower bound in D.

A function $f : D \rightarrow D$ is said to be *monotonic* if $f(x) \leq f(y)$ whenever $x \leq y$. It is said to be (finitely) distributive if its application distributes over finite meets, that is if

$$f(a \sqcap b) = f(a) \sqcap f(b),$$

and it is said to be infinitely distributive if its application distributes over infinite meets, that is if

$$f\left(\underset{a \in A}{\sqcap} a \right) = \underset{a \in A}{\sqcap} f(a),$$

for arbitrary sets A. In dataflow analysis terminology an infinitely distributive function is also called a continuous function, though, elsewhere, a continuous function is defined to be a function that distributes over chains.

Given a set D, we denote the set of all k-tuples of elements from D by D^k. If $(D, \sqcap)$ is a meet semilattice then the partial ordering $\leq$ on D induces a corresponding partial ordering on D^k based on component-wise ordering, and D^k is itself a meet semilattice with respect to this ordering. Further, if D has a top element then so does D^k, and if D is bounded then so is D^k.

We denote the set of all functions from D to D by $D \rightarrow D$. Given a semilattice $(D, \sqcap)$, a meet operation $\sqcap$ is defined on $D \rightarrow D$ as follows:

$$f \sqcap g =_{def} \lambda x. (f(x) \sqcap g(x)).$$

$(D \rightarrow D, \sqcap)$ is a meet semilattice of functions over D. The set of all distributive functions over D will be denoted by $D \rightarrow_d D$. It can be verified that $D \rightarrow_d D$ is closed with respect to function composition and the meet operation. Hence, $(D \rightarrow_d D, \sqcap)$ is itself a meet semilattice.

Let Q be a collection of k equations in k variables x_1 through x_k, the i-th equation being

$$x_i = g_i(x_1, \ldots, x_k),$$

where the variables range over values from a partially ordered set $(D, \leq)$. A *fixed point* of this collection of equations is a k-tuple of values satisfying the collection of equations. If the set of all fixed points of Q has a greatest element (with respect to the partial ordering on D^k), this greatest fixed point, also called the maximum fixed point of Q, will be denoted by $MFP_D(Q)$. The subscript D will be omitted if no confusion is likely.

Note that the k equations above can be combined into one equation

$$X = G(X),$$

where X ranges over D^k, and $G : D^k \rightarrow D^k$ combines the functions g_1 through g_k. (If each g_i is monotonic then G itself will be monotonic.) Hence, computing the maximum fixed point of a collection of equations is, in some sense, no more difficult than computing the maximum fixed point of a single equation. It is well known that if G is monotonic and D^k is bounded and has a greatest element $\top$, then the maximum fixed point of the above equation is given by the limit of the descending chain $\top \geq G(\top) \geq G(G(\top)) \geq \cdots G^i(\top) \cdots$. (See, for example, [Sto77].)

5.3. Path Problems in Graphs

5.3.1. Algebraic Path Problems

The *algebraic path problem*, a generalization of the shortest-path problem, has attracted wide attention, principally because a large variety of problems can be formulated as algebraic path problems and algorithms developed for the shortest-path problem can be generalized to solve these problems. This algebraic generalization was introduced by Carre [Car71]. Some subsequent surveys of this problem and extensions of Carre's work may be found in [Zim81], [Mah84], [Gon84], [Gon84a], and [Rot90].

An *algebraic path problem* is specified by a set S and two binary operators on S: a "path extension" operator $\otimes$ and a "summary" operator $\oplus$, both of which satisfy certain properties to be discussed soon. An *instance* of this algebraic path problem is given by a directed graph $G = (V,E)$ and an edge-labelling function l that associates a value $l(e)$ from S with each edge $e \in E$.

We first discuss a way of formulating the algebraic path problem suggested by the straightforward way of formulating the shortest-path problem. Consider an instance (G,l) of the problem. The function l can be extended to map paths in G to elements of S as follows. The *label* of a path $p = [e_1, e_2, \ldots, e_k]$ is defined by $l(p) = l(e_1) \otimes l(e_2) \otimes \cdots \otimes l(e_k)$. If v, w are two vertices in the graph, then $C(v,w)$ is defined to be the *summary* over all paths p from v to w of $l(p)$:

$$C(v,w) = \bigoplus_{p : v \rightarrow^* w} l(p).$$

(We use the notation $p : v \rightarrow^* w$ to denote the set of all paths p from v to w.)

For the classical shortest-path problem the set S is the set of reals, the operator $\otimes$ is addition, while the operator $\oplus$ is "min". Just like in the shortest-path problem, we can define several variants of the algebraic path problem, such as the "all-pairs" variant, the "single-source" variant, and the "single-pair" variant. In an all-pairs problem, the goal is to compute $C(v,w)$ for all pairs of vertices $v, w \in V(G)$. In a single-source problem, the goal is to compute the value $C(s,w)$ for each vertex w, where s is a distinguished source vertex. In the single-pair problem, only a specific value $C(s,t)$ is to be computed, where s and t are given vertices.

Note that the above definition of the label of a path makes sense only if $\otimes$ is an associative operator. The definition of $C(v,w)$ also makes sense only under certain assumptions about the operator $\oplus$. Since the order in which the labels of paths are to be "summed" up is not specified, $\oplus$ has to be an associative and commutative operator. Since the set of paths between two vertices in a graph may be infinite there is also the question of what an infinite "summary" means.

Before addressing the issues raised above, let us first examine a second way of formulating the shortest-path problem, which suggests an alternative formulation of the algebraic path problem that avoids the concept of summary over an infinite set.

Every input instance G of the single-source shortest-path problem (SSoSP) induces a collection of equations, the *Bellman-Ford* equations, in the set of unknowns $\{ x_u \mid u \in V(G) \}$:

$$x_u = \min(\{0\} \cup \{ [x_v + length(v \rightarrow u)] \mid v \in Pred(u) \}) \quad \text{if } u = source(G)$$
$$= \min_{v \in Pred(u)} [x_v + length(v \rightarrow u)] \qquad\qquad \text{otherwise.}$$

The maximum fixed point of this collection of equations is the solution to the SSoSP problem if the input graph contains no negative length cycles. This suggests the idea of defining the algebraic path problem as that of the computation of the maximum fixed point of the following collection Q_P of equations determined by an input instance $P = (G, source, l)$:

$$x_u = \oplus(\{\bar{1}\} \cup \{ [x_v \otimes l(v \rightarrow u)] \mid v \in Pred(u) \}) \quad \text{if } u = source(G)$$
$$= \bigoplus_{v \in Pred(u)} [x_v \otimes l(v \rightarrow u)] \qquad\qquad \text{otherwise,}$$

where $\bar{1}$ is the identity element with respect to $\otimes$. This problem definition does not make any assumptions about the applicability of the summary operation to infinite sets, but it does assume the existence of a partial ordering among the values so that one may talk of the maximum fixed point.

We now review the definition of various algebraic systems utilized in defining various classes of algebraic path problems, consider the properties required for formulating the problem in either of the ways outlined above, and study the equivalence of the above two definitions under some conditions.

Definition 5.1. A *semiring* (also known as a *dioid*) is a system $(S, \oplus, \otimes, \bar{0}, \bar{1})$ consisting of a set S, two binary operations $\oplus$ and $\otimes$ on S, and two elements $\bar{0}$ and $\bar{1}$ of S, satisfying the following axioms:

1. $(S, \oplus, \bar{0})$ is a commutative monoid:

$(a)\, x \oplus y = y \oplus x$	(commutativity)
$(b)\, x \oplus (y \oplus z) = (x \oplus y) \oplus z$	(associativity)
$(c)\, x \oplus \bar{0} = x$	(identity)

2. $(S, \otimes, \bar{1})$ is a monoid:

$(a)\, x \otimes (y \otimes z) = (x \otimes y) \otimes z$	(associativity)
$(b)\, x \otimes \bar{1} = x = \bar{1} \otimes x$	(identity)

3. $\otimes$ distributes over $\oplus$:

$(a)\, x \otimes (y \oplus z) = (x \otimes y) \oplus (x \otimes z)$	(right-distributivity)
$(b)\, (x \oplus y) \otimes z = (x \otimes z) \oplus (y \otimes z)$	(left-distributivity)

4. $\bar{0}$ is an annihilator with respect to $\otimes$:

$x \otimes \bar{0} = \bar{0} = \bar{0} \otimes x$	(annihilation)

Observation 5.2. As the name suggests, semirings are closely related to rings. A *ring (with unity)* is a semiring that satisfies the following additional axiom:
$$\forall x\ \exists y.\ x \oplus y = \bar{0} \qquad\qquad\qquad \text{(invertibility)}$$
The definition of a ring does not normally include the annihilation axiom since it follows as a logical consequence of the remaining axioms including invertibility.

Observation 5.3. Let $(S,\ \oplus,\ \otimes,\ \bar{1})$ be a system that satisfies axioms $1(a)$, $1(b)$, 2 and 3 of Definition 5.1. Let $\bar{0}$ denote an element not in S. Then, $(S \cup \{\bar{0}\},\ \oplus,\ \otimes, \bar{0},\ \bar{1})$ is a semiring, where the operators $\oplus$ and $\otimes$ are extended to handle $\bar{0}$ according to the identity axiom $1(c)$ and annihilation axiom 4. Hence, the requirement of a zero element that satisfies axioms $1(c)$ and 4 is not crucial to the definition of a semiring. Note that axiom 3 implies that $\otimes$ distributes over finite but non-empty summations of the form $x_1 \oplus x_2 \oplus \cdots \oplus x_k$. Axiom 4 can be viewed as the requirement of distributivity over the empty sum too, since the empty sum can naturally be defined to yield the identity element $\bar{0}$.

Let $\mathcal{R}$ denote the set of reals and let $\mathcal{R}^{\geq 0}$ denote the set of non-negative reals. Let $+$ and $\times$ denote the operations of addition and multiplication on reals, and let min denote the minimum operator on reals. Both $(\mathcal{R},\ +,\ \times,\ 0,\ 1)$ and $(\mathcal{R} \cup \{\infty\},\ \min,\ +,\ \infty,\ 0)$ are examples of semirings. The operations of the latter semiring have to be extended to handle ∞ in the obvious way, and the inclusion of ∞ in the latter semiring illustrates Observation 5.3.

Let $S = (S,\ \oplus,\ \otimes, \bar{0}, \bar{1})$ be a specific semiring. We now define the SS-S problem (the single-source problem determined by the semiring S), adopting the first of the two possible formulations of the problem discussed earlier.

Definition 5.4. Given a directed graph $G = (V, E)$, a vertex s in V, and an edge-labeling function $l : E \rightarrow S$, the *SS-S problem* is to compute $C(s,w)$ for every vertex w in V. We say that (G, s, l) is an *instance* of the SS-S problem.

For example, let S be the semiring $(\mathcal{R}^{\geq 0} \cup \{\infty\},\ \min,\ +,\ \infty,\ 0)$. Then, SS-$S$ is nothing other than the single-source shortest-path problem with non-negative edge lengths. If S is the semiring $(\{0,1\}, \max, \min, 0, 1)$, then SS-$S$ is the single-source reachability problem. The corresponding all-pairs version of the problem is the transitive closure problem. Hence, algebraic path problems are sometimes also known as *generalized transitive closure* problems.

As we pointed out earlier, the definition of $C(s,w)$ may involve the "summary" of an infinite set of values, since there may be infinitely many paths between two vertices in a cyclic graph. Unfortunately, the semiring axioms are not sufficient to extend the summary operator to infinite sets. We now consider some special semirings in which it is possible to extend the summary operator of a semiring to handle infinite sets.

Definition 5.5. A *sum-ordered semiring* is a semiring $(S,\ \oplus,\ \otimes,\ \bar{0},\ \bar{1})$ in which the relation $\leq$ defined by
$$x \leq z \text{ iff there exists an } y \text{ such that } z = x \oplus y \qquad\qquad \text{(sum-ordering)}$$

is a partial ordering.

For example, $(\mathcal{R}^{\geq 0} \cup \{\infty\}, +, \times, 0, 1)$ is a sum-ordered semiring. In this case the sum-ordering defined above is the standard total ordering on reals. We now consider an important class of sum-ordered semirings.

Definition 5.6. An *idempotent semiring* is a semiring $(S, \oplus, \otimes, \overline{0}, \overline{1})$ that satisfies the following addition axiom:

$$x \oplus x = x \qquad\qquad\qquad\qquad \text{(idempotence)}$$

Observation. The $\oplus$ operator of an idempotent semiring satisfies the axioms of a *semilattice*. In other words, the binary relation $\leq$ defined by:

$$x \leq y \text{ iff } x \oplus y = x$$

is a partial ordering and $\oplus$ is the meet or greatest-lower-bound operator with respect to this partial ordering. It can be verified easily that an idempotent semiring is also a sum-ordered semiring, though the ordering as defined above is the inverse of the sum-ordering.

But if the $\oplus$ operator is a greatest-lower-bound operator, then we no longer have any problem *defining* the concept of the summary of an infinite set of values. The notion of a greatest-lower-bound applies even to infinite sets, though the greatest-lower-bound of an infinite set need not *exist* in general. The greatest-lower-bound of a set X of elements, when it exists, will be denoted by $\bigoplus_{x \in X} x$. Consider an instance of the SS-S problem, where S is an idempotent semiring. For every vertex u, either $C(s,u)$ does not exist or is a precisely defined quantity. For example, consider the idempotent semiring $(\mathcal{R} \cup \{\infty\}, \min, +, \infty, 0)$, which defines the single-source shortest-path problem. If there exists a path from s to u that passes through a negative-length cycle, then one can construct paths from s to u of arbitrarily small length. Consequently, the set of lengths of all paths from s to u has no greatest-lower-bound, and $C(s,u)$ does not exist.[2] In every other case, $C(s,u)$ is precisely defined.

We can adopt this idea and extend the concept of summary of infinite sets to sum-ordered semirings too. Given an infinite set X, we can define the sum of X to be the least-upper-bound of the set of all "partial sums" of finite subsets of X: $\bigoplus X =_{def} \bigsqcup \{\bigoplus Y : Y$ is a finite subset of $X\}$. For $(\mathcal{R}^{\geq 0} \cup \{\infty\}, +, \times, 0, 1)$, which is sum-ordered but not idempotent, this yields a definition of infinite summation that is equivalent to the standard notion of infinite summation of non-negative reals as the limit of a sequence of partial sums. An example of an algebraic path problem involving this semiring is that of computing the stationary probabilities of a Markov process: we have a graph with a source vertex, and every edge has an associated value between 0 and 1, which is the transition probability associated with that edge. Consequently, the value associated with a path is the probability that the transition sequence

---

[2]If we add $-\infty$ to the domain, then the greatest-lower-bound *would* exist in this case, and equal $-\infty$. Since defining $\infty + (-\infty)$ is tricky, we would have to let $+$ be a partial operator.

indicated by that path occurs. The value $C(s,u)$ indicates the probability of arriving at vertex u from vertex s.

We now turn to the problem of *computing* the solution to an instance of a semiring problem. Though we may be able to meaningfully define what an infinite summary means, we still face the problem of how one computes an infinite summary. This also brings us to the second way of looking at algebraic path problems. We now look at a special class of idempotent semirings where the two approaches to algebraic path problems are equivalent.

Definition 5.7. A *closed semiring* is an idempotent semiring $(S, \oplus, \otimes, \bar{0}, \bar{1})$ satisfying the following additional axioms:

(a) Every countable set X has a greatest lower bound $\displaystyle\bigoplus_{x \in X} x$ (completeness)

$(b)\; \left(\displaystyle\bigoplus_{x \in X} x \right) \otimes \left(\displaystyle\bigoplus_{y \in Y} y \right) = \displaystyle\bigoplus_{(x,y) \in X \times Y} x \otimes y$ (infinite distributivity)

The axioms satisfied by a closed semiring enable us to show that the solution $\{ C(s,u) \mid u$ is a vertex $\}$ to an instance $P=(G,s,l)$ of the single-source closed semiring problem is the maximum fixed point of the set of equations Q_P induced by the input instance of the problem. They also make it possible to use an adaptation of the Floyd-Warshall algorithm for the shortest-path problem [Flo62] to solve the all-pairs version of an instance of the closed semiring problem. Well, almost. We still need a way of computing infinite summaries. In using the adaptation of the Floyd-Warshall algorithm to the closed semiring problem, it is usually assumed that implementations of the $\oplus$ and $\otimes$ operators are available, as well as the operator * defined below.

Definition 5.8. A unary operator *, called **closure**, of a closed semiring $(S, \oplus, \otimes, \bar{0}, \bar{1})$ is defined as follows:

$$a^* =_{def} \bigoplus_{i=0}^{\infty} a^i$$

where $a^0 = \bar{1}$ and $a^{i+1} = a^i \otimes a$.

Let us now consider a further special class of semirings.

Definition 5.9. An idempotent semiring $(S, \oplus, \otimes, \bar{0}, \bar{1})$ is said to be a *Dijkstra semiring* if the underlying partial ordering is actually a total ordering, and if $\bar{1}$ is the least element with respect to this ordering.

Hence, a Dijkstra semiring essentially consists of a totally ordered set S with a least element $\bar{1}$ and a greatest element $\bar{0}$, and an operator $\otimes$ that is monotonic with respect to the total ordering such that $(S, \otimes, \bar{1})$ is a monoid. The name "Dijkstra semiring" follows from the fact that an adaptation of Dijkstra's algorithm can be used to find the solution to an instance of the SS-S problem whenever S is a Dijkstra semiring. The identity $a^* = \bar{1}$ holds in a Dijkstra semiring; hence, an implementation of the closure operator is not necessary to compute the solution to an instance of the Dijkstra semiring problem, unlike in the case of the closed semiring problem.

5.3.2. Dataflow Analysis Problems

A lattice-theoretic framework for dataflow analysis problems was first developed by Kildall [Kil73] and later refined by Kam and Ullman [Kam76, Kam77] and Graham and Wegman [Gra76]. In this framework, the program to be analyzed is represented by a *flowgraph* G with a distinguished *entry* vertex s. The problem requires the computation of some information $S(u)$ for each vertex u in the flowgraph. In general, this information describes the possible state of the program whenever execution reaches the program point u. More precisely, $S(u)$ summarizes the set of all states that can arise at point u. The values $S(u)$ are assumed to be elements of a meet semilattice L. The interpretation of the meet operation is that if $a \in L$ summarizes a set of states S_1, and $b \in L$ summarizes a set of states S_2, then $a \sqcap b$ summarizes the set of states $S_1 \cup S_2$. A value $c \in L$ is associated with the entry vertex which describes the program state that can exist at the entry vertex when the program begins execution. A (monotonic) function $M(e):L \rightarrow L$ is associated with every edge e in the flowgraph: the interpretation of this function is that $M(u \rightarrow v)(a)$ describes (summarizes) the program state at point v when after program execution reaches point v along the edge $u \rightarrow v$, given that the program state at point u is described by a. (The above informal explanation of the various elements involved is typically formalized via the notion of abstract interpretation. See, for example, [Abr87, Cou77].)

The function M can be extended to map every path in the graph to a function from L to L: if p is a path $[e_1, e_2, \ldots, e_k]$ then $M(p)$ is defined to be $M(e_k) \circ M(e_{k-1}) \cdots M(e_1)$. The desired (meet-over-all-paths) solution $S(u)$ is defined as follows:

$$S(u) = \underset{P:s \rightarrow^* u}{\sqcap} M(p)(c)$$

Recall that $p : s \rightarrow^* u$ denotes that the meet is taken over the set of all paths p from s to u.

A semilattice $(L, \sqcap)$, a set of functions $F \subseteq L \rightarrow L$ and a constant $c \in L$, often the greatest or least element of the semilattice, constitute a *dataflow analysis framework*. We will say that the dataflow analysis framework is bounded if L is bounded, that it is closed if L is closed with respect to arbitrary meets, and that it is monotonic, distributive, or infinitely distributive if every function in F is respectively monotonic, distributive, or infinitely distributive.

Every dataflow analysis framework $\mathcal{F} = (L, \sqcap, F, c)$ determines a specific dataflow analysis problem, which we call the $\mathcal{F}$-DFA problem. An input instance of the problem consists of a graph G with an entry vertex s and a mapping M from the edges of G to F. Such an input instance induces a collection of equations:

$$S(s) = \sqcap(\{c\} \cup \{M(v \rightarrow s)(S(v)) \mid v \rightarrow s \in E(G)\}$$
$$S(u) = \underset{v \rightarrow u \in E(G)}{\sqcap} M(v \rightarrow u)(S(v)), \quad \text{for } u \neq s.$$

An algorithm for the *(bounded) distributive dataflow analysis problem* is an algorithm that can solve the $\mathcal{F}$-DFA problem for any (bounded) distributive dataflow analysis framework $\mathcal{F}$. Kildall [Kil73] presents an algorithm for the bounded distributive dataflow analysis problem. In particular, he shows that the meet-over-all-paths solution is given by the maximum fixed point of the collection of equations induced by the input instance whenever the framework is distributive. Thus, if the framework is bounded too, then the maximum fixed point can be computed by iteration, as explained earlier.

Kam and Ullman show that the similarly defined *bounded monotonic dataflow analysis problem* is undecidable: that is, they show that no algorithm exists that can solve the $\mathcal{F}$-DFA problem for every bounded monotonic framework $\mathcal{F}$. They also show for monotonic frameworks that the maximum fixed point of the above equations is less than or equal to the meet-over-all-paths solution. Thus, when the framework is bounded, the maximum fixed point is a computable "approximation" to the desired solution.

5.3.3. The Relation between Dataflow Analysis and Algebraic Path Problems

We now study the relationship between the dataflow analysis framework and the algebraic path problem framework.

Theorem 5.10. Corresponding to every idempotent semiring S there exists a distributive dataflow analysis framework $\mathcal{F}$ such that the SS-S problem and $\mathcal{F}$-DFA problem are equivalent. Further, if S is bounded so is $\mathcal{F}$, and if S is closed then $\mathcal{F}$ is infinitely distributive and closed (with respect to meets).

Proof.

Let $(S, \oplus, \otimes, \bar{0}, \bar{1})$ be the given idempotent semiring S. For every $a \in S$ let f_a denote the function $\lambda x.x \otimes a$. Let $\mathcal{F}$ be the dataflow analysis framework $(S, \oplus, F, \bar{1})$, where F is the set of functions $\{ f_a \mid a \in S \}$. $\mathcal{F}$ is distributive since for any f_a in F, $f_a(x \oplus y) = (x \oplus y) \otimes a = (x \otimes a) \oplus (y \otimes a) = f_a(x) \oplus f_a(y)$. Similarly, it follows that if S is closed then $\mathcal{F}$ is infinitely distributive (continuous) and closed with respect to meets.

The SS-S problem and $\mathcal{F}$-DFA problem are equivalent in the following sense: there is an isomorphism between the input instances of the two problems such that the corresponding input instances have the same solutions; any input instance of one of the problems can be trivially transformed into an input instance of the other problem whose solution readily yields the solution to the original problem instance. More specifically, an instance (G,s,l) of SS-S problem corresponds to an instance (G,s,M) of the $\mathcal{F}$-DFA problem, where the mapping M is defined by $M(e) = \lambda x.x \otimes l(e)$. $\square$

Theorem 5.11. Corresponding to every distributive dataflow analysis framework $\mathcal{F}$ there exists an idempotent semiring S such that the $\mathcal{F}$-DFA problem reduces to the SS-S problem.

Proof.

Let $(L, \sqcap, F, c)$ be the given dataflow analysis framework $\mathcal{F}$. Assume, without loss of generality, that L has a greatest element $\top$. For every $a \in L$, define the constant-valued function $g_a : L \rightarrow L$ by $g_a(x) = a$. Note that in a dataflow analysis framework we have a set of values L and a set of functions F, while in a semiring we have only a set of values. We will embed both L and F into a semilattice of functions over L by identifying every element a in L with the constant-valued function g_a. Let i denote the identity function. Define S to be the smallest set of functions containing F $\cup \{ g_a \mid a \in L \} \cup \{ i \}$ that is closed under the meet operation.

We now show that $S = (S, \sqcap, \circ, g_\top, i)$ is an idempotent semiring. Since $\sqcap$ is a semilattice operator it satisfies the axioms of commutativity, associativity and idempotence, and $g_\top$ is the identity with respect to $\sqcap$. Function composition is associative with i as the identity element. We now verify that $\circ$ distributes over $\sqcap$: $f_1 \circ (f_2 \sqcap f_3)$ $= \lambda x. f_1((f_2 \sqcap f_3)(x)) = \lambda x. f_1(f_2(x) \sqcap f_3(x)) = \lambda x. f_1(f_2(x)) \sqcap f_1(f_3(x)) = \lambda x. (f_1 \circ f_2)(x) \sqcap (f_1 \circ f_3)(x) = (f_1 \circ f_2) \sqcap (f_1 \circ f_3)$. Left-distributivity can be verified similarly. Now, $g_\top$ may not satisfy the annihilator axiom, but that is not significant. As explained in Observation 5.3, we can introduce a new annihilator element that will make S a semiring.

We now show that the $\mathcal{F}$-DFA problem reduces to the SS-S problem. We can transform every instance (G, s, M) of the $\mathcal{F}$-DFA problem into an instance (G', s', l) of the SS-S problem as follows. The graph G' is obtained by adding a new source vertex s' and a new edge $s' \rightarrow s$ to the graph G. The mapping l is defined by $l(e) = M(e)$ for every edge e in the original graph, and $l(s' \rightarrow s) = g_c$. (The solution to this transformed input instance associates every vertex with the constant-valued function g_a, where a is the value associated with that vertex in the solution to the original input instance.)

Not every instance of the SS-S problem can be translated back into an instance of the $\mathcal{F}$-DFA problem, and, in this sense, the $\mathcal{F}$-DFA and SS-S problems may not be equivalent. However, if the set of functions F contains all constant-valued functions and the identity function and is closed with respect to the meet operation, then $S = F$, and the $\mathcal{F}$-DFA and SS-S problems *are* equivalent. $\square$

The above theorems establish the equivalence of dataflow analysis problems and algebraic path problems in the following sense: an algorithm that can solve every closed semiring problem can be trivially adapted to solve every infinitely distributive dataflow analysis problem and vice versa.

5.4. Grammar Problems

5.4.1. The Idea Behind The Problem

Knuth defined the following generalization of the single-source shortest-path problem [Knu77]: Consider a context free grammar in which every production is associated with a real-valued function whose arity equals the number of non-terminal occurrences on the right-hand side of the production. Every derivation of a terminal string from a non-terminal has an associated derivation tree; replacing every production in the derivation tree by the function associated with that production yields an expression tree. Define the cost of a derivation to be the value of the expression tree obtained from the derivation. The problem is to compute for each non-terminal of the given grammar the minimum-cost derivation of a terminal string from that non-terminal.

Example 5.12. ([Knu77]). The following example illustrates the above definition. We will consider how an instance of the single-source shortest-path problem can be transformed into an equivalent instance of the grammar problem. Given an instance of the SSoSP problem, define a grammar consisting of one non-terminal N_u for every vertex u in the given graph. For every edge $u \rightarrow v$ in the graph, we add a new terminal $g_{u \rightarrow v}$, and a production $N_v \rightarrow g_{u \rightarrow v}(N_u)$ associated with the production function f defined by $f(x) = x + length(u \rightarrow v)$. In addition, we add the production $N_s \rightarrow 0$, where s is the source vertex and 0 is a terminal and the production function is the constant-valued function zero. Now, there is a bijective correspondence between the set of all paths from the source vertex to a vertex v and the set of terminal strings derivable from N_v. Further, the "costs" of corresponding paths and terminal strings are equal. Hence, the solution to the grammar problem instance is the same as the solution to the original SSoSP instance.

Thus, the single-source shortest-path problem (with non-negative edge lengths) corresponds to the special case of the grammar problem where the input grammar is regular and all the production functions g are of the form $g(x) = x + h$ (for some $h \geq 0$) or $g() = 0$. Further, the grammar problem corresponds to an SSoSP problem only if contains exactly one production of the form $N \rightarrow 0$; if more than one production is of this form, then we have a "simultaneous multi-source shortest-path problem".

Knuth showed that it is possible to adapt Dijkstra's shortest-path algorithm [Dij59] to solve the grammar problem if the functions defining the costs of derivations satisfy a simple property. In addition to the single-source shortest-path problem, Knuth lists a variety of other applications and special cases of the grammar problem, including the generation of optimal code for expression trees and the construction of optimal binary-search trees.

It is possible to generalize Knuth's problem by borrowing ideas from the algebraic path problem and the dataflow analysis problem. For example, it is useful to consider cases where the value associated with a terminal string or a derivation is not a real value but some value drawn from a partially ordered set, and the value to be computed is the meet over the set of all values associated with the terminal strings derivable from a non-terminal. For instance, define the value of a non-empty terminal string to be a set consisting of the first character in the string, and define the value of the empty string to be $\{\varepsilon\}$. Define the summary operation to be set union. This gives us the first-set computation problem (where the set computed for a non-terminal contains ε iff that non-terminal can derive the empty string).

Before we define this generalized problem, let us consider another application that motivates it—interprocedural dataflow analysis.

In intraprocedural dataflow analysis—dataflow analysis of programs without procedure calls—the program to be analyzed is represented by a control-flow graph with an entry vertex. Every path p from the entry vertex to a vertex u has an associated value, which provides information about the program state that would exist at point u if program execution were to follow path p. The meet over all the paths from the entry to u of the value associated with that path yields information about the program state that can exist at point u irrespective of the path that program execution takes to reach u. Note that a path p from the entry vertex to u may be dynamically infeasible—that is, there may exist no input state for which program execution will follow that path under the standard model of execution. But it is, in general, undecidable if a given path in the graph is dynamically feasible. Consequently, it is common in dataflow analysis to summarize the set of statically feasible paths rather than the set of dynamically feasible paths, where a statically feasible path may be formalized as follows: branches in a control-flow graph are typically governed by predicates, whose evaluation determines which of the branches should be taken; consider a non-standard model of execution in which the value of every such predicate is determined randomly; we will refer to a path program execution can take under this model of execution as a statically feasible path. For a single-procedure control-flow graph every path is a statically feasible path, and, hence, the concept of a statically feasible path seems to serve no purpose. But this concept proves to be relevant in interprocedural dataflow analysis.

Interprocedural dataflow analysis concerns the dataflow analysis of programs with multiple procedures. We first consider how the flow of control in a program with procedures can be represented by the multi-procedure control-flow graph. Each procedure p is represented by a single-procedure control-flow graph that has a special entry vertex p_{entry} and a special exit vertex p_{exit}. We assume that each call-site c of a procedure p is represented by two vertices in the control-flow graph—c_{pre}, which represents the point just before the procedure is invoked, and c_{post}, which represents the point just after the procedure call returns. There is no edge from c_{pre} to c_{post}— instead, there is an edge from c_{pre} to p_{entry}, and an edge from p_{exit} to c_{post}, to represent the transfer of program control during procedure invocation and return. We will refer to these two edges as matching interprocedural edges. Now, every possible path that program execution can take is represented by a path in this graph. But, not every path from the entry vertex of the graph is statically feasible. In a statically feasible path, interprocedural edges have to be matched, much like matching parentheses in a well-formed expression. It would be desirable in interprocedural dataflow analysis to compute the meet over all statically feasible paths, instead of the meet over all paths. But, how do we characterize statically feasible paths?

Context-free grammars provide a convenient mechanism for characterizing statically feasible paths in a multi-procedure control-flow graph. (This idea appears in [Sha81] and in [Bin91].) Given a multi-procedure control-flow graph, we can represent it by a context free grammar such that the set of all terminal strings derivable from a given non-terminal represents the set of all statically feasible paths between two specific program points. The set of non-terminals N, the set of terminals T, and the set of productions P of this grammar are defined as follows:

$N = \{ p(u,v) \mid u \text{ and } v \text{ are vertices } \}$

$T = \{ e(u,v) \mid u \longrightarrow v \text{ is an edge in the multi-procedure control-flow graph } \}$

$P = \{ p(u,u) \longrightarrow \varepsilon \mid u \text{ is a vertex } \} \cup$

$\quad \{ p(u,w) \longrightarrow p(u,v)e(v,w) \mid u, v, w \text{ are vertices, and }$

$\quad\quad\quad\quad\quad\quad\quad v \longrightarrow w \text{ is a non-return edge } \} \cup$

$\quad \{ p(u,c_{post}) \longrightarrow p(u,c_{pre})e(c_{pre},q_{entry})p(q_{entry},q_{exit})e(q_{exit},c_{post}) \mid$

$\quad\quad\quad\quad u \text{ is a vertex, and } c \text{ is a call to procedure } q \}.$

It can be easily verified that the terminal strings derivable from a non-terminal $p(u,v) \in N$ do describe statically feasible paths from u to v.

This formalization of the set of statically feasible paths via context free grammars suggests that the grammar problem may provide a convenient mechanism for formalizing interprocedural dataflow analysis. We will see in the next section that this is the case.

5.4.2. The Problem Definition

We now define the generalized version of Knuth's problem. Unlike the algebraic path problem, this problem makes use only of a summary operator. The role of the extension operator is taken over by that of "production functions". In all the problems we consider in this section the summary operator will be the meet or join operator over a partially ordered set. The problem Knuth studies in [Knu77] concerns the totally ordered set $(\mathcal{R}^{\geq 0} \cup \{\infty\}, \min)$. The following definition is an extension of Knuth's formulation to arbitrary partially ordered sets.

Definition 5.13. An *abstract grammar* over a semilattice $(D, \sqcap)$ is a context free grammar in which all productions are of the general form

$$Y \longrightarrow g(X_1, \ldots, X_k),$$

where $Y, X_1, \ldots, X_k$ are non-terminal symbols, and g, the parentheses and commas are all terminal symbols. In addition, each production $Y \longrightarrow g(X_1, \ldots, X_k)$ has an associated function from D^k to D, which will be denoted by g itself in order to avoid the introduction of more notation. The function g is referred to as a *production function* of the given grammar.

We adopt the restriction that each production in an abstract grammar have the special form $Y \longrightarrow g(X_1, \ldots, X_k)$ to maintain continuity with respect to the notation used by Knuth and to simplify notation. Observe that the special form of the productions ensures that the grammar is non-ambiguous. The terminal strings generated by such productions can be thought of as describing the parse trees or abstract syntax trees of some other context free grammar. Thus, an abstract grammar is essentially an attribute grammar with productions of this form, where each nonterminal has a single synthesized attribute.

For every non-terminal symbol Y of an abstract grammar over the terminal alphabet T we let $L(Y) = \{ \alpha \mid \alpha \in T^* \text{ and } Y \longrightarrow^* \alpha \}$ be the set of terminal strings derivable from Y. Every string α in $L(Y)$ denotes a composition of production functions, so it corresponds to a uniquely defined value in D, which we shall call *val(α)*.

Given a semilattice $(D, \sqcap)$, the *abstract grammar problem* is to compute the value $m_G(Y)$ for each non-terminal Y of a given abstract grammar G over $(D, \sqcap)$, where

$$m_G(Y) =_{def} \bigsqcap_{\alpha \in L_G(Y)} val(\alpha).$$

We will drop the subscript in $m_G(Y)$ if no confusion is likely.

Example 5.14. ([Knu77]). Let us look at some simple examples of the grammar problem over the semilattice $(\mathcal{R}^{\geq 0} \cup \{\infty\}, \min)$. Given a context free grammar, consider the abstract grammar obtained by replacing each production $Y \rightarrow \theta$ in the given grammar by the production $Y \rightarrow g_\theta(X_1, \ldots, X_k)$, where $X_1, \ldots, X_k$ are the non-terminal symbols occurring in θ from left to right (including repetitions). If we define the production function g_θ by

$$g_\theta(x_1, \ldots, x_k) =_{def} x_1 + \cdots + x_k + (\textit{the number of terminal symbols in } \theta)$$

then $m_G(Y)$, the solution to the resulting grammar problem, is the length of the shortest terminal string derivable from non-terminal Y. If we instead define g_θ by

$$g_\theta(x_1, \ldots, x_k) =_{def} max(x_1, \ldots, x_k) + 1$$

then $m_G(Y)$ is the minimum height of a parse tree for a string derivable from the non-terminal Y. The nullable non-terminals problem can similarly be expressed as a grammar problem over the semilattice $(\{0, 1\}, \min)$ by defining g_θ as follows:

$$g_\theta(x_1, \ldots, x_k) =_{def} max(x_1, \ldots, x_k) \qquad \text{if } \theta \text{ has no terminal symbols}$$
$$=_{def} 1 \qquad \text{otherwise}$$

Note that as a special case of the above definition $g_\theta()$ is 0 if θ is the empty string. It follows that $m_G(Y)$ is 0 if Y can derive the null string, and 1 otherwise. $\square$

Example 5.15. Let us now see how the first-set computation problem can be expressed as a grammar problem. Given a context free grammar C, with a set of non-terminals N and a set of terminals T, we want to compute for every non-terminal Y in N the set of all terminals a such that Y can derive a terminal string beginning with a. To simplify the presentation we assume, without loss of generality, that every production in C is either of the form $Y \rightarrow X_1 \cdots X_k$ where every symbol on the right-hand side is a non-terminal or of the form $Y \rightarrow a$ where a is either a terminal or the empty string ε. We construct an abstract grammar G over the semilattice $(2^T, \cup)$ with N as the set of non-terminals as follows. Corresponding to every production $Y \rightarrow a$ in C, G has a production $Y \rightarrow g()$ where the production function g is defined by $g()=\{a\}$. Corresponding to every production $Y \rightarrow X_1 \cdots X_k$, G has a production $Y \rightarrow g(X_1, \ldots, X_k)$, where the production function g is defined by:

$$g(s_1, \ldots, s_k) = \left(\bigcup_{i=1}^{k} t_i \right) \cup e,$$

where t_i = **if** s_j contains ε for every $j < i$ **then** $s_i - \{\varepsilon\}$ **else** $\emptyset$,
and e = **if** every s_j contains ε **then** $\{\varepsilon\}$ **else** $\emptyset$.

Now, every terminal string α of G corresponds to a terminal string $\overline{\alpha}$ of the original grammar C, and it can be verified that $val(\alpha)$ is the set containing the first terminal in $\overline{\alpha}$. $\square$

The above definition of the abstract grammar problem was motivated by Knuth's problem. The meet-over-all-paths formulation of dataflow analysis problems, however, suggests a variant of the grammar problem in which every terminal string of the grammar is associated with a *function*. In particular, we define the *functional grammar problem* over a semilattice $(D, \sqcap)$ and a constant $c \in D$ as follows. An input instance of this problem consists of a context free grammar G and a mapping M from the terminals of G to $D \rightarrow D$. For any terminal string $\alpha = a_1 \cdots a_k$ we define $M(\alpha)$ to be $M(a_k) \circ \cdots \circ M(a_1)$. We then consider the problem of computing the value $\underset{\alpha \in L_G(Y)}{\sqcap} M(\alpha)(c)$ for each non-terminal Y of the given grammar G. In view of the explanation in the previous section, it should be obvious how interprocedural dataflow analysis can be formalized as a functional grammar problem.

Combining ideas from the above two problem formulations leads to the following generalization in which the function $M(\alpha)$ associated with a terminal string α may be defined in more complex ways (than in the functional grammar problem) using production functions.

Given a semilattice $(D, \sqcap)$ and a constant $c \in D$, the *generalized functional grammar problem* is to compute the value $\underset{\alpha \in L_G(Y)}{\sqcap} val(\alpha)(c)$ for each non-terminal Y of a given abstract grammar G over $(D \rightarrow D, \sqcap)$.

It is easily seen that the generalized functional grammar problem is really a generalization of the functional grammar problem. In particular, we can trivially reformulate every functional grammar problem as a generalized functional grammar problem as follows. Given an instance (G, M) of the functional grammar problem, the corresponding instance of the generalized functional grammar problem is the abstract grammar G' over $(D \rightarrow D, \sqcap)$ which, for every production $X \rightarrow \alpha_1 Y_1 \alpha_2 Y_2 \cdots \alpha_k Y_k \alpha_{k+1}$ in G consists of the production $X \rightarrow F(Y_1, \ldots, Y_k)$, where the production function F is given by

$$F(f_1, \ldots, f_k) = M(\alpha_{k+1}) \circ f_k \circ \cdots \circ M(\alpha_2) \circ f_1 \circ M(\alpha_1). \tag{5.1}$$

Similarly, the abstract grammar problem can be easily reduced to the generalized functional grammar problem as follows. Given an abstract grammar G over the semilattice $(D, \sqcap)$, consider the abstract grammar G' over the semilattice $(D \rightarrow D, \sqcap)$. For every production $X \rightarrow g(Y_1, \ldots, Y_k)$ in G, G' contains a production $X \rightarrow F(Y_1, \ldots, Y_k)$, where the production function F is defined by

$$F(f_1, \ldots, f_k) = \lambda x. g(f_1(x), \ldots, f_k(x)).$$

Unfortunately, we cannot formulate the functional grammar problem (or the generalized functional grammar problem) as an abstract grammar problem. Note that we are interested in computing $\underset{\alpha \in L_G(Y)}{\sqcap} (val(\alpha)(c)) = (\underset{\alpha \in L_G(Y)}{\sqcap} val(\alpha))(c)$ in the generalized functional grammar problem. The problem of computing the value $\underset{\alpha \in L_G(Y)}{\sqcap} val(\alpha)$ is directly an abstract grammar problem over a semilattice of functions. This is important because we will soon see how the abstract grammar problem can be expressed as a maximum fixed point problem, which can be solved using standard techniques, provided that the semilattice is bounded.

5.4.3. The Grammar Problem as a Maximum Fixed Point Problem

The grammar problem for an abstract grammar G naturally leads to a collection of mutually recursive equations, which consists of the following equation for each non-terminal Y in the grammar.

$$f(Y) = \underset{Y \to g(X_1, \ldots, X_k)}{\sqcap} g(f(X_1), \ldots, f(X_k))$$

We will refer to this collection of equations as Q_G, or more briefly Q. Note that Q contains one variable, $f(Y)$, for each non-terminal Y in the grammar. Consequently, a solution or fixed point of Q consists of a value $v(Y)$ for each non-terminal Y. A tuple of such values will be denoted by $(v(Y) \mid Y$ is a non-terminal$)$ or just by v.

We now show that the solution m_G to the grammar problem is the maximum fixed point of Q_G, provided each production function of the abstract grammar G satisfies a simple property.

Definition 5.16. Let $(D, \sqcap)$ be a meet semilattice. A function $g : D^k \to D$ is said to be a *finitely distributive function* if for every collection of k non-empty finite (index) sets $I_1, \ldots, I_k$,

$$g\left(\underset{i_1 \in I_1}{\sqcap} x_{i_1}, \ldots, \underset{i_k \in I_k}{\sqcap} x_{i_k}\right) = \underset{(i_1, \ldots, i_k) \in I_1 \times \cdots \times I_k}{\sqcap} g(x_{i_1}, \ldots, x_{i_k}) \quad (5.2)$$

in the sense that if the expression on either side has a well-defined value, the expression on the other side is well-defined too and has the same value. The function is said to be *weakly distributive* if it satisfies the above property for arbitrary non-empty index sets. and it is said to be *infinitely distributive* if it satisfies the above property for arbitrary index sets.

Note that the production function defined in Example 5.15 is infinitely distributive. Now, consider the production function F (see equation (5.1)) utilized to encode the functional grammar problem as a generalized functional grammar problem. This function is infinitely distributive over the semilattice $(D \to_d D, \sqcap)$ if all the constant functions $M(\alpha_i)$ occurring on the right-hand side of the equation are distributive, that is, if the function $M(a)$ associated with any terminal symbol a is distributive.

Let $\sqcap$ be the meet or join operation of a totally ordered set D. It can be easily verified that a function from D^k to D is finitely distributive iff it is monotonic. If $\sqcap$ is the meet operation of a bounded semilattice, then the notions of weak distributivity and finite distributivity coincide.

In general, when $\sqcap$ is the meet operation of a partially ordered set D, every finitely distributive function is monotonic, but a monotonic function need not be finitely distributive. A weakly distributive function g is not required to satisfy equation (5.2) if any of the sets $I_1, \ldots, I_k$ is empty. Let $\top$ denote the top element of D. Obviously, a weakly distributive function is infinitely distributive iff $g(x_1, \ldots, x_k)$ is $\top$ whenever any x_i is $\top$.

We now consider the collection of equations Q. We say an abstract grammar is a *distributive grammar* if every production function of the grammar is distributive.

Monotonic grammars are similarly defined.

Lemma 5.17. If G is an infinitely distributive grammar then $(m_G(Y) \mid Y$ is a non-terminal) is a fixed point of Q_G.

Proof.

$$m_G(Y) = \bigcap_{Y \to^{\cdot} \alpha} val(\alpha) \qquad (\text{ from the definition of } m_G(Y))$$

$$= \bigcap_{Y \to g(X_1, \ldots, X_k)} \bigcap_{g(X_1, \ldots, X_k) \to^{\cdot} \alpha} val(\alpha)$$

$$= \bigcap_{Y \to g(X_1, \ldots, X_k)} \bigcap \{val(g(\alpha_1, \ldots, \alpha_k)) \mid X_i \to^{\cdot} \alpha_i\}$$

$$= \bigcap_{Y \to g(X_1, \ldots, X_k)} \bigcap \{g(val(\alpha_1), \ldots, val(\alpha_k)) \mid X_i \to^{\cdot} \alpha_i\}$$

$$(\text{ from the definition of } val(g(\alpha_1, \ldots, \alpha_k)))$$

$$= \bigcap_{Y \to g(X_1, \ldots, X_k)} g(\bigcap_{X_1 \to^{\cdot} \alpha_1} val(\alpha_1), \ldots, \bigcap_{X_k \to^{\cdot} \alpha_k} val(\alpha_k))$$

$$(\text{ since } g \text{ is infinitely distributive})$$

$$= \bigcap_{Y \to g(X_1, \ldots, X_k)} g(m_G(X_1), \ldots, m_G(X_k))$$

$$(\text{ from the definition of } m_G)$$

□

Lemma 5.18. Let G be a monotonic grammar, and let $(f(Y) \mid Y$ is a non-terminal) be a fixed point of Q_G. Then, $f(Y) \leq m_G(Y)$ for each non-terminal Y.

Proof. It is sufficient to show for every terminal string α that if Y is a non-terminal such that $Y \to^{\cdot} \alpha$, then $f(Y) \leq val(\alpha)$. The proof is by induction on the length of the string α. Assume $Y \to^{\cdot} \alpha$. Then we must have $Y \to g(X_1, \ldots, X_k) \to^{\cdot} g(\alpha_1, \ldots, \alpha_k) = \alpha$. Since each α_i is a smaller string than α and $X_i \to^{\cdot} \alpha_i$, it follows from the inductive hypothesis that $f(X_i) \leq val(\alpha_i)$. It follows from the monotonicity of g that $g(f(X_1), \ldots, f(X_k)) \leq g(val(\alpha_1), \ldots, val(\alpha_k)) = val(\alpha)$. Since $(f(Y) \mid Y$ is a non-terminal) is a fixed point of Q we have $f(Y) \leq g(f(X_1), \ldots, f(X_k))$. The result follows. □

The above lemma generalizes the result of Kam and Ullman [Kam77] relating the maximum fixed point (MFP) and the meet-over-all-paths (MOP) solutions to a monotone dataflow analysis problem, since every monotone dataflow analysis problem is isomorphic to a monotone grammar problem.

Theorem 5.19. Let G be a infinitely distributive grammar. Then $(m_G(Y) \mid Y$ is a non-terminal) is the maximum fixed point of Q_G.

Proof. Immediate from lemmas 5.17 and 5.18. □

The above theorem generalizes Kildall's result [Kil73] that the MFP solution yields the MOP solution for (infinitely) distributive dataflow analysis problems and also a similar result for interprocedural analysis due to [Sha81]. The above result shows how grammar problems can be reduced to maximum fixed point computation problems. An interesting feature of the grammar problem is that it is equivalent to maximum fixed point computation problems, in the following sense: maximum fixed point computation problems can be reduced to grammar problems, under some distributivity assumptions. Let us say that a collection of equations is infinitely distributive if the function on the right-hand side of each equation is infinitely distributive. For instance, the collection of equations Q_G determined by an infinitely distributive grammar G is infinitely distributive. Now, given an infinitely distributive collection of equations Q, it is easy to construct an infinitely distributive grammar whose solution is the maximum fixed point of the given collection of equations, as follows. For every variable x_i, introduce a non-terminal N_i and a production $N_i \rightarrow \infty$, where ∞ is a terminal, associated with the constant-valued production function ∞; for every equation $x_i = g_i(x_1, \ldots, x_k)$, introduce a new production $N_i \rightarrow g_i(N_1, \ldots, N_k)$. It can be easily verified that the collection of equations induced by this grammar trivially simplifies to the collection Q we started with. Consequently, the problem of computing the maximum fixed point of an infinitely distributive collection of equations is equivalent to the infinitely distributive grammar problem.

Now, the maximum fixed point of the collection of equations Q_G can be computed by iteration if $(D, \sqcap)$ is a bounded semilattice. Thus, the above theorem yields an effective algorithm for the distributive grammar problem over bounded semilattices. What does this say about interprocedural dataflow analysis? We observed earlier that a distributive interprocedural dataflow analysis problem over a semilattice $(D, \sqcap)$ can be reduced to a distributive grammar problem over the semilattice of functions $(D \rightarrow _dD, \sqcap)$. Consequently, such an interprocedural analysis problem can be solved using techniques for finding maximum fixed points provided that $D \rightarrow _dD$ itself is a bounded semilattice. Since computing the maximum fixed point of a collection of equations involves evaluating the right-hand side of the equations, we also need to be able to compute the composition and meet of functions easily. (Solving intraprocedural dataflow analysis problems using elimination techniques also requires us to be able to compute the composition and meet of functions; thus, dataflow analysis problems solvable using elimination techniques typically meet this requirement.) $D \rightarrow _dD$ can be unbounded even if D is bounded. However, $D \rightarrow _dD$ will be bounded if D is finite. Thus, $(Gen, Kill)$ type of dataflow analysis problems (see [Aho86] or [Fis88], for instance), where the functions involved can be represented by a pair of finite sets Gen and $Kill$, can typically be solved using the above technique. This approach to solving interprocedural dataflow analysis problems is, however, not new—it appears in [Sha81].

5.4.4. The SSF Grammar Problem and the SWSF Fixed Point Problem

In the previous two sections we looked at the generalization of Knuth's problem motivated by the algebraic path problem and dataflow analysis. In this section we return to problems that are closer to Knuth's problem. From the point of view of incremental computation, the general grammar problem is difficult. Since it generalizes the algebraic path problem and the dataflow analysis problem, it is, in fact, unbounded. (See Chapter 8.) However, we present an efficient and bounded incremental algorithm for a variant of Knuth's problem in the next chapter. The goal of this section is to define and study this problem. In this section we assume that the semilattice $(D, \sqcap)$ is a totally ordered set—hence, the meet operation is really the min operation.

Let $[i,k]$ denote the set of integers $\{ j \mid i \leq j \leq k \}$. A function $g(x_1, \ldots, x_k)$ from D^k to D is said to be a *superior function* (abbreviated s.f.) if it is monotone non-decreasing in each variable and if $g(x_1, \ldots, x_k) \geq x_i$ for every $i \in [1,k]$ and for every $x_1, \ldots, x_k$. A function $g(x_1, \ldots, x_k)$ from D^k to D is said to be a *strict superior function* (abbreviated s.s.f.) if it is monotone non-decreasing in each variable and if $g(x_1, \ldots, x_k) > x_i$ for every $i \in [1,k]$. An abstract grammar in which every production function is a superior function is said to be an *SF grammar*. An abstract grammar in which every production function is a strict superior function is said to be an *SSF grammar*. Examples of superior functions over $(\mathcal{R}^{\geq 0}, \leq, \infty)$ include $max(x_1, \ldots, x_k)$, $x+y$, and $\sqrt{x^2 + y^2}$. None of these functions are strict superior functions over the set of non-negative reals, although the latter two are strict superior functions over the set of positive reals. Every 0-ary function, that is, a function that has no input arguments, is trivially an s.s.f. It can be verified that the production functions used in Example 5.14 are all superior functions. Consequently, the abstract grammars defined there are all SF grammars.

The motivation for the above definitions come from the shortest-path problem. Recall that when a shortest-path problem is encoded as a grammar problem (see Example 5.12), the production functions used are of the form $g(x) = x + h$, where h is the length of an edge, or $g() = 0$. The function g defined by $g(x) = x + h$ is an s.f. function if $h \geq 0$ and an s.s.f. function if $h > 0$. Consequently, the abstract grammar generated by an instance of the SSoSP≥ 0 problem is an SF grammar, while the abstract grammar generated by an instance of the SSoSP>0 problem is an SSF grammar. Knuth shows that Dijsktra's algorithm for the SSoSP≥ 0 problem can be generalized to solve the SF grammar problem. Similarly, the incremental algorithm we presented in Chapter 4 for the SSoSP>0 problem can be generalized to solve the dynamic SSF grammar problem. We will address the dynamic SSF grammar problem in the next chapter.

Let us now consider the fixed point formulation of these problems. We now define two classes of functions that generalize the class of superior and strict superior functions respectively, which will be utilized in defining the fixed point problem. We say a function $g : D^k \rightarrow D$ is a *weakly superior function* (abbreviated w.s.f.) if it is monotone non-decreasing in each variable and if for every $i \in [1,k]$,

$$g(x_1, \ldots, x_i, \ldots, x_k) < x_i \;\Rightarrow\; g(x_1, \ldots, x_i, \ldots, x_k) = g(x_1, \ldots, \infty, \ldots, x_k).$$

We say a function $g : D^k \rightarrow D$ is a *strict weakly superior function* (abbreviated s.w.s.f.) if it is monotone non-decreasing in each variable and if for every $i \in [1,k]$,

$$g(x_1, \ldots, x_i, \ldots, x_k) \leq x_i \;\Rightarrow\; g(x_1, \ldots, x_i, \ldots, x_k) = g(x_1, \ldots, \infty, \ldots, x_k).$$

It can be easily verified that every s.f. is also a w.s.f., while every s.s.f. is also an s.w.s.f. The function $min(x_1, \ldots, x_k)$ is an example of a w.s.f. that is not an s.f., while $min(x_1, \ldots, x_k)+1$ is an example of an s.w.s.f. that is not an s.s.f. A constant-valued function is another example of an *s.w.s.f.*

Now consider a collection Q of k equations in the k unknowns x_1 through x_k, the i-th equation being

$$x_i = g_i(x_1, \ldots, x_k). \qquad (\dagger)$$

An equation of this form is said to be a WSF equation if g_i is a *w.s.f.*, and an SWSF equation if g_i is an *s.w.s.f.* Note that the expression on the right-hand side of the equation need not contain all the variables and that it may be more precisely written as

$$x_i = g_i(x_{j_{i,1}}, x_{j_{i,2}}, \ldots, x_{j_{i,n(i)}}).$$

We will continue to use the earlier form of the equation as a notational convenience.

The motivation for the above definitions comes from the type of equations generated by an SF or SSF grammar. Recall that every instance of a grammar problem generates a collection of equations consisting of the following equation for each non-terminal Y in the grammar:

$$d(Y) = \min \{ g(d(X_1), \ldots, d(X_k)) \mid Y \rightarrow g(X_1, \ldots, X_k) \text{ is a production} \}.$$

It can be shown (see below) that if every production function is a *w.s.f.* then the above equation is a WSF equation. This is because min is also a *w.s.f.* function, and *w.s.f.* functions are closed under composition. It follows that every equation generated by an instance of the SF grammar problem is a WSF equation. Similarly, every equation generated by an instance of the SSF grammar problem is an SWSF equation.

It can be shown that if each of the equations in Q is a WSF equation then Q has a maximum fixed point, which can be computed by an adaptation of Dijkstra's algorithm. We define the *WSF maximum fixed point problem* to be that of computing the maximum fixed point of a collection of WSF equations. This problem generalizes the SF grammar problem.

However, we are interested in incremental algorithms for the fixed point problem, and it turns out to be necessary to address a restricted version of the WSF maximal fixed point problem. If each of the equations in Q is an SWSF equation, then Q can be shown to have a unique fixed point. (See below.) We define the *SWSF fixed point problem* to be that of computing the unique fixed point of a collection of SWSF equations. The SWSF fixed point problem generalizes the SSF grammar problem, since each equation in the collection of equations determined by an SSF grammar is an SWSF equation, as we show later. (The SSoSP>0 problem is obtained as yet a further special case of the SSF grammar problem; that is, when all edge lengths are positive, the Bellman-Ford equations are all SWSF.)

We now establish the various claims made above. We first establish some properties of *s.w.s.f.* functions that will be useful later on. Thinking about an *s.w.s.f.* of the form $min(x_1+h_1, \ldots, x_k+h_k)$, where each $h_i > 0$ may make it easier to understand the proposition.

Proposition 5.20.

(a) Let $g : D^k \rightarrow D$ be a *s.w.s.f.* and let $I \subseteq \{1, \ldots, k\}$ be such that $g(x_1, \ldots, x_k)$
      $\leq x_i$ for every $i \in I$. Then,

$$g(y_1, \ldots, y_k) = g(x_1, \ldots, x_k)$$

where $y_i =_{def}$ if $(i \in I)$ then ∞ else x_i.

(b) Let $g : D^k \rightarrow D$ be a s.w.s.f. and let $x_1, \ldots,$ x_k be such that $g(x_1, \ldots, x_i, \ldots, x_k) \le x_i$. Then,

(1) For all $y \ge g(x_1, \ldots, x_i, \ldots, x_k)$,

$$g(x_1, \ldots, y, \ldots, x_k) = g(x_1, \ldots, x_i, \ldots, x_k)$$

(2) For all $y < g(x_1, \ldots, x_i, \ldots, x_k)$,

$$g(x_1, \ldots, y, \ldots, x_k) > y$$

(c) If g is a s.w.s.f. and $g(x_1, \ldots, x_k) < g(y_1, \ldots, y_k)$ then there exists $i \in [1,k]$ such that $x_i < g(x_1, \ldots, x_k)$ and $x_i < y_i$.

(d) If g is a s.w.s.f. and $g(x_1, \ldots, x_i, \ldots, x_k) \ne g(x_1, \ldots, x_i', \ldots, x_k)$, then $g(x_1, \ldots, x_i, \ldots, x_k) > min(x_i, x_i')$ and, similarly, $g(x_1, \ldots, x_i', \ldots, x_k) > min(x_i, x_i')$.

Proof.

(a) This follows by repeated applications of the definition of an s.w.s.f.

(b) Let $x_1, \ldots, x_k$ be such that $g(x_1, \ldots, x_i, \ldots, x_k) \le x_i$. We now prove (1). Let $y \ge g(x_1, \ldots, x_i, \ldots, x_k)$. We show that $g(x_1, \ldots, y, \ldots, x_k) = g(x_1, \ldots, x_i, \ldots, x_k)$ by assuming otherwise and deriving a contradiction.

$$g(x_1, \ldots, y, \ldots, x_k) \ne g(x_1, \ldots, x_i, \ldots, x_k)$$

$\Rightarrow g(x_1, \ldots, y, \ldots, x_k) \ne g(x_1, \ldots, \infty, \ldots, x_k)$ (since g isan s.w.s.f.)

$\Rightarrow g(x_1, \ldots, y, \ldots, x_k) < g(x_1, \ldots, \infty, \ldots, x_k)$ (since g is monotonic)

$\Rightarrow g(x_1, \ldots, y, \ldots, x_k) < g(x_1, \ldots, x_i, \ldots, x_k)$ (since g is an s.w.s.f.)

$\Rightarrow g(x_1, \ldots, y, \ldots, x_k) < y$ (from assumption about y)

$\Rightarrow g(x_1, \ldots, y, \ldots, x_k) = g(x_1, \ldots, \infty, \ldots, x_k)$ (since g is an s.w.s.f.)

$\Rightarrow g(x_1, \ldots, y, \ldots, x_k) = g(x_1, \ldots, x_i, \ldots, x_k)$ (since g is an s.w.s.f.)

The result follows. Now (2) follows as a simple consequence of (1). Suppose there exists some $y < g(x_1, \ldots, x_i, \ldots, x_k) \le x_i$ such that $g(x_1, \ldots, y, \ldots, x_k) \le y$. Thus, we have $g(x_1, \ldots, y, \ldots, x_k) \le y$ and $x_i \ge g(x_1, \ldots, y, \ldots, x_k)$. Using (1), but with the roles of x_i and y reversed, we have $g(x_1, \ldots, x_i, \ldots, x_k) = g(x_1, \ldots, y, \ldots, x_k) \le y$, which is a contradiction.

(c) We prove the contrapositive. Assume that the conclusion is false. Hence, for every $x_i < g(x_1, \ldots, x_k)$ we have $x_i \ge y_i$. Then,

$$g(x_1, \ldots, x_k) = g(z_1, \ldots, z_k)$$

where $z_i =_{def}$ if $(x_i \ge g(x_1, \ldots, x_k))$ then ∞ else x_i

(from (a))

$\ge g(y_1, \ldots, y_k)$ since every $z_i \ge y_i$.

(since g is monotonic)

The result follows.

(d) This follows directly from (b), since if $g(x_1, \ldots, x_i, \ldots, x_k) \le x_i$, then $g(x_1, \ldots, x_i, \ldots, x_k) = g(x_1, \ldots, y, \ldots, x_k)$ for all $y \ge g(x_1, \ldots, x_i, \ldots, x_k)$. $\square$

We now show that the class of w.s.f. and s.w.s.f. functions are closed with respect to function composition.

Proposition 5.21.
    (a) If $g(x_1, \ldots, x_k)$ is a s.w.s.f. then so is the function $h(x_1, \ldots, x_m)$ defined by

$$h(x_1, \ldots, x_m) =_{def} g(x_{j_1}, \ldots, x_{j_k})$$

where every $j_i \in [1, m]$. Similarly, if g is a w.s.f. then so is h.
    (b) Let $f(x_1, \ldots, x_k)$ be a w.s.f., and let $g_j(x_1, \ldots, x_m)$ be a s.w.s.f. for every $j \in [1, k]$. The function $h(x_1, \ldots, x_m)$ defined as follows is a s.w.s.f. too.

$$h(x_1, \ldots, x_m) =_{def} f(g_1(x_1, \ldots, x_m), \ldots, g_k(x_1, \ldots, x_m))$$

Further, if each g_j is a w.s.f., then so is h.

Proof.
    (a) Let g be a s.w.s.f. The monotonicity of g directly implies that h is monotonic. Now,

$$h(x_1, \ldots, x_m) \leq x_i$$
$$\Rightarrow g(x_{j_1}, \ldots, x_{j_k}) \leq x_i$$
$$\Rightarrow g(x_{j_1}, \ldots, x_{j_k}) \leq x_{j_p} \quad \text{for every } p \text{ such that } j_p = i$$
$$\Rightarrow g(y_1, \ldots, y_k) = g(x_{j_1}, \ldots, x_{j_k}) \text{ where } y_p =_{def} \text{ if } (j_p = i) \text{ then } \infty \text{ else } x_{j_p}$$
$$\qquad \text{using Proposition 5.20}(a)$$
$$\Rightarrow h(x_1, \ldots, \infty, \ldots, x_m) = h(x_1, \ldots, x_i, \ldots, x_m)$$

It similarly follows that if g is a w.s.f. then h is a w.s.f. too.
    (b) The monotonicity of h follows immediately from the monotonicity of f and $g_1, \ldots, g_k$. Now,

$$h(x_1, \ldots, x_i, \ldots, x_k) \leq x_i$$
$$\Rightarrow f(y_1, \ldots, y_k) \leq x_i \quad \text{where } y_j =_{def} g_j(x_1, \ldots, x_i, \ldots, x_k)$$
$$\Rightarrow f(y_1, \ldots, y_k) < y_j \quad \text{for every } y_j > x_i$$
$$\Rightarrow f(w_1, \ldots, w_k) = f(y_1, \ldots, y_k) \quad \text{where } w_j =_{def} \text{ if } (y_j > x_i) \text{ then } \infty \text{ else } y_j$$
$$\qquad \text{(using Proposition 5.20}(a))$$
$$\Rightarrow f(w_1, \ldots, w_k) = f(y_1, \ldots, y_k) \text{ where}$$
$$\qquad w_j =_{def} \text{ if } (g_j(x_1, \ldots, x_i, \ldots, x_k) > x_i) \text{ then } \infty \text{ else } g_j(x_1, \ldots, x_i, \ldots, x_k)$$
$$\qquad = \text{ if } (g_j(x_1, \ldots, x_i, \ldots, x_k) > x_i) \text{ then } \infty \text{ else } g_j(x_1, \ldots, \infty, \ldots, x_k)$$
$$\qquad\qquad \text{since } g_j \text{ is strictly weakly superior}$$
$$\qquad \geq g_j(x_1, \ldots, \infty, \ldots, x_k)$$
$$\Rightarrow f(z_1, \ldots, z_k) \leq f(y_1, \ldots, y_k) \text{ where } z_j =_{def} g_j(x_1, \ldots, \infty, \ldots, x_k)$$
$$\Rightarrow h(x_1, \ldots, \infty, \ldots, x_k) \leq h(x_1, \ldots, x_i, \ldots, x_k)$$
$$\Rightarrow h(x_1, \ldots, \infty, \ldots, x_k) = h(x_1, \ldots, x_i, \ldots, x_k)$$
$$\qquad \text{since } h(x_1, \ldots, \infty, \ldots, x_k) \geq h(x_1, \ldots, x_i, \ldots, x_k) \text{ by monotonicity}$$

The result follows. $\square$

    We now characterize the set of equations determined by SF and SSF grammars.

Theorem 5.22. If G is a SF grammar, then Q_G is a collection of WSF equations, while if G is an SSF grammar, then Q_G is a collection of SWSF equations.

Proof. Every equation in Q_G is of the form
$$d(Y) = \min (g_1(d(X_{i_{1,1}}), \ldots, d(X_{i_{1,n(1)}})), \ldots, g_m(d(X_{i_{m,1}}), \ldots, d(X_{i_{m,n(m)}}))).$$
Now, *min* is a *w.s.f.* It follows from Proposition 5.21 that if each g_i is an *s.f.* then the above equation is an WSF equations (since a superior function is also a weakly superior function). Similarly, if each g_i is an *s.s.f.*, then the above equation is an *s.w.s.f.* (since an *s.s.f.* is also an *s.w.s.f.*). The result follows. □

Theorem 5.23. Let Q be a collection of k equations, the i-th equation being
$$x_i = g_i(x_1, \ldots, x_k).$$
If every g_i is an *s.w.s.f.* then Q has a unique fixed point.

Proof. The existence of a fixed point will follow from the algorithm to be presented in the next Chapter, which computes this fixed point. The uniqueness of the fixed point may be established as follows.

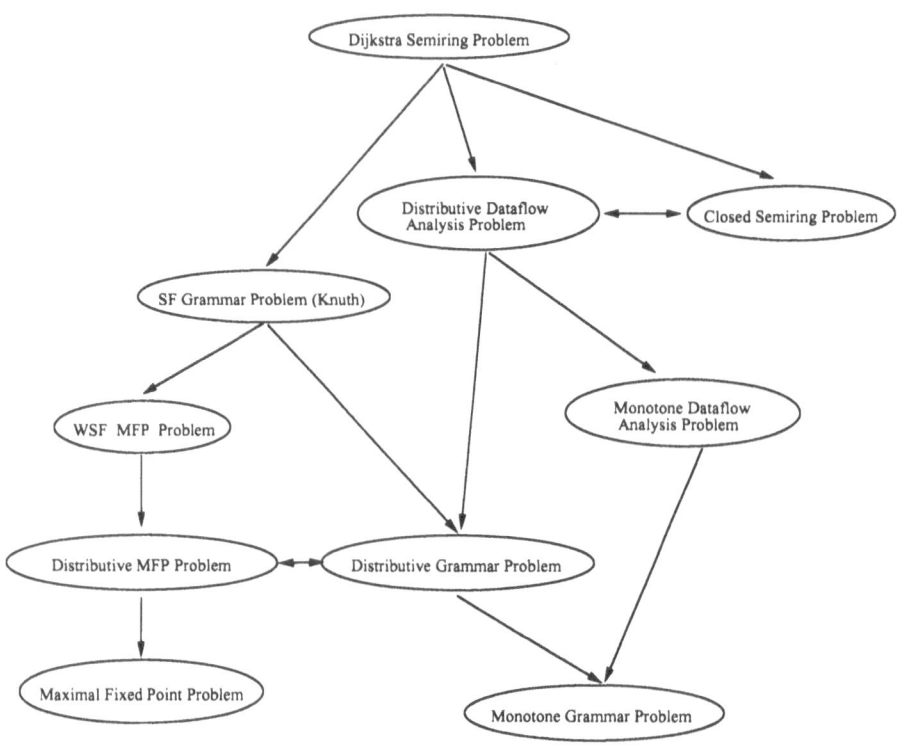

Figure 5.1. The above figure illustrates the relationship among different classes of problems. Every ellipse represents a class of problems and an arrow represents containment—more precisely, an arrow from one class of problems to another implies that every problem in the first class reduces to some problem in the second class. Not all classes of problems considered in this chapter are depicted in the above picture.

Assume, to the contrary, that $(a_i \mid 1 \leq i \leq k)$ and $(b_i \mid 1 \leq i \leq k)$ are two different fixed points of Q. Choose the least element of the set $\{ a_i \mid a_i \neq b_i \} \cup \{ b_i \mid a_i \neq b_i \}$. Without loss of generality, assume that the least element is a_i. Thus, we have $a_i < b_i$, and also $a_j = b_j$ for all $a_j < a_i$. Now, we derive a contradiction as follows.

$$a_i = g_i(a_1, \ldots, a_k) \quad \text{since}(a_i \mid 1 \leq i \leq k) \text{ is a fixed point of } Q$$
$$= g_i(c_1, \ldots, c_k) \quad \text{where } c_j =_{def} \text{ if } (a_j < a_i) \text{ then } a_j \text{ else } \infty$$
$$\text{(since } g_i \text{ is a strict } w.s.f.)$$
$$= g_i(c_1, \ldots, c_k) \quad \text{where } c_j =_{def} \text{ if } (a_j < a_i) \text{ then } b_j \text{ else } \infty$$
$$\text{(since } a_j = b_j \text{ whenever } a_j < a_i)$$
$$\geq g_i(b_1, \ldots, b_k) \quad \text{since } c_j \geq b_j \text{ for every } j \in [1,k]$$
$$\geq b_i \quad \text{since } (b_i \mid 1 \leq i \leq k) \text{ is a fixed point of } Q.$$

The contradiction implies that Q has a unique fixed point. $\square$

It is worth mentioning at this point that the above results hold in somewhat more general form. Define a WSF grammar to be an abstract grammar in which every production function is a $w.s.f.$, and an SWSF grammar to be an abstract grammar in which every production function is an $s.w.s.f.$ The grammar problem for a WSF grammar that has no useless symbols—a context free grammar is said to have no useless symbols if each non-terminal in the grammar can derive at least one terminal string—can be solved by reducing it to the WSF maximum fixed point problem. It is straightforward to show that, conversely, the WSF maximum fixed point problem can be reduced to the grammar problem for a WSF grammar with no useless symbols. Similarly, the grammar problem for an SWSF grammar with no useless symbols is equivalent to the SWSF fixed point problem.

5.5. Related Work

In this chapter we have looked at the algebraic path problem and the dataflow analysis problem and at generalizations of these problems. Figure 5.1 illustrates the relationship between several classes of problems considered in this chapter.

Lengauer et al. [Len91] explore possible generalizations of the algebraic path problem motivated by examples where some of the semiring axioms fail to hold. A problem similar to the grammar problem arises in the context of query evaluation in deductive databases, where derived relations may be defined using logic programs. Database query languages provide aggregate operations such as *max, min, count,* and *sum,* which may be applied to a selected field of a relation. These aggregate operations lead to the problem of "summarizing" sets defined via logic programs. Sudarshan [Sud92] discusses special cases of this problem where techniques similar to those used in Dijkstra's shortest-path algorithm can be used.

Chapter 6
An Incremental Algorithm for a Generalization
of the Shortest-Path Problem

When it is not necessary to change, it is necessary not to change.
—Lord Falkland

6.1. Introduction

In this chapter we present an algorithm for the dynamic SWSF fixed point problem. (The material presented in this chapter also appears in [Ram].) The algorithm updates the (unique) fixed point of a collection of SWSF equations after an arbitrary (possibly non-unit) change to the collection of equations. We then present an improved version of this algorithm, adapted for the dynamic SSF grammar problem. As a special case of the algorithm, we obtain a new, simple, and efficient algorithm for the dynamic single-source shortest-path problem with positive edge lengths (the *dynamic SSSP>0 problem*).

The aspect of the algorithm we present in this chapter that distinguishes it from all other work on dynamic shortest-path problems, including our own work described in Chapter 4, is that it handles *multiple heterogeneous changes*: Between updates, the input graph is allowed to be restructured by an arbitrary mixture of edge insertions, edge deletions, and edge-length changes. Most previous work on dynamic shortest-path problems has addressed the problem of updating the solution after the input graph undergoes either *unit changes*—i.e, exactly one edge is inserted, deleted, or changed in length—or else *homogeneous changes*—i.e., changes to multiple edges are permitted, but all changes must be of the same kind: either all insertions/length-decreases or all deletions/length-increases. (A comprehensive comparison of our work with previous work appears in Section 6.6.)

In general, a single application of an algorithm for heterogeneous changes has the potential to perform significantly better than either the repeated application of an algorithm for unit changes or the double application of an algorithm for homogeneous changes. There are two sources of potential savings: *combining* and *cancellation*.

Combining: If updating is carried out by using multiple applications of an algorithm for unit or homogeneous changes, a vertex might be examined several times, with the vertex being assigned a new (but temporary and non-final) value on each visit until the last one. An algorithm for heterogeneous changes has the potential to combine the effects of all of the different modifications to the input graph, thereby eliminating the extra vertex examinations.

Cancellation: The effects of insertions and deletions can cancel each other out. Thus, if updating is carried out by using multiple applications of an algorithm for unit or homogeneous changes, superfluous work can be performed. In one updating pass, vertices can be given new values only to have a subsequent updating pass revisit the vertices, restoring their original values. With an algorithm for heterogeneous changes, there is the potential to avoid such needless work.

The updating algorithm presented in this chapter exploits these sources of potential savings to an essentially optimal degree: if the initial value of a vertex is already its correct, final value, then the value of that vertex is never changed during the updating; if the initial value of a vertex is incorrect, then either the value of the vertex is changed only once, when it is assigned its correct final value, or the value of the vertex is changed exactly twice, once when the value is temporarily changed to ∞, and once when it is assigned its correct, final value. (Bear in mind that, when updating begins, it is not known which vertices have correct values and which do not.)

As a consequence, the incremental algorithm we present for the shortest-path problem is a bounded one. In particular, the algorithm updates the shortest-path solution in time $O(\|\delta\| \log \|\delta\|)$, after an arbitrary mixture of edge insertions, edge deletions, and edge-length changes.[1] The incremental algorithm we present for the SWSF grammar problem is a bounded cost scheduling algorithm.

Though the algorithms presented in this chapter are *incremental* algorithms, they can be seen as generalizations of *batch* algorithms. For instance, Dijkstra's algorithm turns out to be a special case of our algorithm for the dynamic SSSP>0 problem: when a collection of edges is inserted into an empty graph, our algorithm works like Dijkstra's algorithm. Similarly, a variant of Knuth's algorithm for the batch grammar problem is obtained as a special case of our algorithm for the dynamic grammar problem. However, our incremental algorithms encounter "configurations" that can never occur in any run of the batch algorithms. For example, in the dynamic SSSP>0 algorithm, a vertex u can, at some stage, have a distance $d(u)$ that is *strictly less than* $\min_{v \in Pred(u)} [d(v) + length(v \rightarrow u)]$. This situation never occurs in Dijkstra's algorithm.

This chapter is organized as follows. In Section 6.2 we present the basic idea behind the algorithm and prove its correctness. We present the first version of our algorithm, a proof of its correctness, and an analysis of its time complexity in Section 6.3. In Section 6.4, we discuss an improved version of the first algorithm, and analyze its time complexity. In Section 6.5 we look at some extensions of the algorithm. In Section 6.6 we discuss related work.

6.2. The Idea Behind the Algorithm

The SWSF fixed point problem (see Section 5.4.4) is to compute the unique fixed point of a collection of SWSF equations. Assume that the given collection of equations consists of k equations in the k unknowns x_1 through x_k, the i-th equation being

$$x_i = g_i(x_1, \ldots, x_k).$$

---

[1] The algorithm can, in fact, update the solution in the same time even if the source (or sink) vertex is changed to be some other vertex in the graph. However, presumably, more vertices will be affected by such a change in the input, and $\|\delta\|$ will be correspondingly larger in this case.

The expression on the right-hand side of the i-th equation need not contain all the variables and it may be more precisely written as

$$x_i = g_i(x_{j_{i,1}}, x_{j_{i,2}}, \ldots, x_{j_{i,n(i)}}).$$

We will continue to use the earlier form of the equation as a notational convenience although an algorithm to compute the fixed point of the collection of equations can use the *sparsity* of the equations to its advantage. We define the *dependence graph* of the collection Q of equations to be the graph (V,E) where $V = \{ x_i \mid 1 \le i \le k \}$, and $E = \{ x_j \rightarrow x_i \mid x_j$ occurs in the right-hand-side expression of the equation for $x_i \}$. For example, the dependence graph of the Bellman-Ford equations induced by an instance (G,s,l) of the single-source shortest-path problem is the graph G itself (with a self-loop for the source vertex s).

For the sake of brevity we will often not distinguish between the collection of equations and the corresponding dependence graph. For instance, we will refer to the variable x_i as "vertex x_i". For convenience, we will refer to the function associated with a vertex x_i by both g_i and g_{x_i}. Every vertex x_i has an associated tentative output value $d[x_i]$, which denotes the value of x_i in the unique fixed point of the collection of equations before modification. Thus, it is the previous output value of vertex x_i. (We use square brackets, as in $d[x_i]$, to indicate variables whose values are maintained by the program.) Let $d^*(x_i)$ denote the actual output value that vertex x_i should have in the unique fixed point of the modified collection of equations. Most of the following terminology is relative to a given assignment d. The *rhs* value of a vertex x_i, denoted by $rhs(x_i)$, is defined to be $g_i(d[x_1], \ldots, d[x_k])$—it denotes the value of the right-hand side of the equation associated with the variable x_i under the given assignment of values to variables. We say that vertex x_i is *consistent* if

$$d[x_i] = rhs(x_i).$$

and that x_i is *inconsistent* otherwise. Two possible types of inconsistency can be identified. We say x_i is an *over-consistent vertex* if

$$d[x_i] > rhs(x_i).$$

We say x_i is an *under-consistent vertex* if

$$d[x_i] < rhs(x_i).$$

A vertex u is said to be a *correct* vertex if $d[u] = d^*(u)$, an *over-estimated* vertex if $d[u] > d^*(u)$, and an *under-estimated* vertex if $d[u] < d^*(u)$. Because $d^*(u)$ is not known for every vertex u during the updating, an algorithm can only make use of information about the "consistency status" of a given vertex, rather than its "correctness status".

We have already seen that the SSSP>0 problem is a special case of the SWSF fixed point problem. Our incremental algorithm for the dynamic SWSF fixed point problem can best be explained as a generalization of Dijkstra's algorithm for the batch shortest-path problem. To draw out the analogy, let us summarize Dijkstra's algorithm using the above terminology.

The collection of equations to be solved in the case of the SSSP>0 problem is the collection of Bellman-Ford equations. In Dijkstra's algorithm all vertices initially have a value of ∞. At any stage of the algorithm, some of the vertices will be consistent while all the remaining vertices will be over-consistent. The algorithm "processes" the inconsistencies in the graph in a particular order: at every stage, it chooses an over-consistent vertex x_i for which the *rhs* value is minimum, and "fixes" this inconsistency by changing $d[x_i]$ to $rhs(x_i)$. The algorithm derives its efficiency by processing the inconsistencies in the "right order", which guarantees that it has to process every vertex at most once.

The idea behind our algorithm is the same, namely to process the inconsistencies in the graph in the right order. The essential difference between our algorithm (for the fully dynamic problem) and Dijkstra's algorithm (for the static problem) is that we need to handle under-consistent vertices as well. Under-consistent vertices can arise in the dynamic shortest-path problem, for instance, when some edge on some shortest path is deleted. This introduces some complications. An inconsistent vertex need not in general be incorrect; an under-consistent vertex need not in general be an under-estimated vertex; and an over-consistent vertex need not in general be an over-estimated vertex. (This is not true in the case of Dijkstra's algorithm, where under-consistent vertices cannot exist, and every overconsistent vertex is guaranteed to be an over-estimated vertex.) See Figure 6.1 for an example illustrating this. *If we change the value of an inconsistent but correct vertex to make it consistent, we may end up with an unbounded algorithm.*

What is the right order for processing inconsistent vertices? We will show that the inconsistencies in the graph should be processed in increasing order of *key*, where the key of an inconsistent vertex x_i, denoted by $key(x_i)$, is defined as follows:

$$key(x_i) =_{def} min(d[x_i], \ rhs(x_i)).$$

In other words, the key of an over-consistent vertex x_i is $rhs(x_i)$, while the key of an

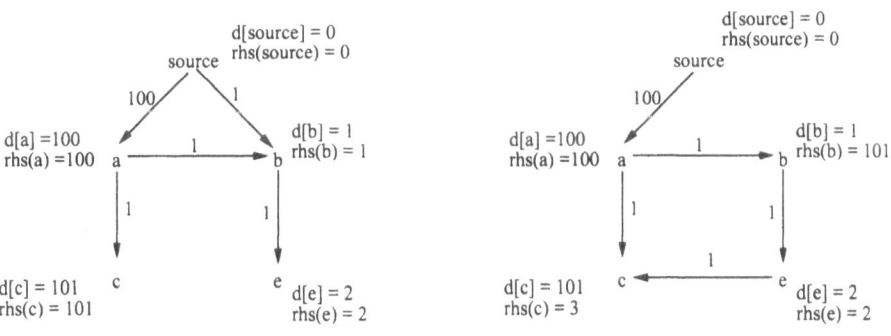

Figure 6.1. Example of overconsistent and underconsistent vertices in the dynamic SSSP>0 problem. The figure on the left indicates a graph for which the single-source shortest-path information has been computed. All vertices are consistent in this graph. The simulataneous deletion of the edge *source* → b and the insertion of the edge e → c make vertex b underconsistent and vertex c overconsistent. Observe that though c is inconsistent it is a correct vertex.

under-consistent vertex x_i is $d[x_i]$. As we will soon show, if u is the inconsistent vertex with the least key, then u is guaranteed to be an over-estimated vertex if it is over-consistent, and it is guaranteed to be an under-estimated vertex if it is under-consistent.

How is an inconsistent vertex to be processed? We will show that if the inconsistent vertex with the least key is over-consistent, then its *rhs* value is its correct value. No such result holds true for under-consistent vertices; however, it turns out that an under-consistent vertex can be "processed" by simply setting its value to ∞, thereby converting it into either a consistent vertex or an over-consistent vertex.

We now present an outline of algorithm in Figure 6.2. The algorithm works by repeatedly selecting an inconsistent variable whose *key* is less than or equal to the key of every other inconsistent variable and processing it. If the selected variable u is under-consistent, then it is assigned a new value of ∞, and if it is over-consistent, then it is assigned the new value of *rhs* (u).

Let us now establish that the algorithm is correct, that it does not change the value of any unaffected vertex, and that it changes the value of a vertex at most twice. We first sketch the idea behind the proof. The above results will follow once we establish the following two claims:

(1) if the vertex u chosen in line [2] is assigned a value in line [6] (in some particular iteration), then vertex u becomes consistent and remains consistent subsequently.

(2) if the vertex u chosen in line [2] is assigned a value in line [4] (in some particular iteration), then u will never be assigned the same value it had before the execution of line [4].

---

procedure IncrementalFP (Q)
declare
        Q : a set of SWSF equations
        $rhs(u) =_{def} g_u(d[x_1], \ldots, d[x_k])$
        $key(u) =_{def} min(d[u], rhs(u))$
begin
[1]   **while** there exist inconsistent variables in Q **do**
[2]   let u be an inconsistent vertex with minimum *key* value
[3]   **if** $d[u] < rhs(u)$ **then**
[4] $d[u] := \infty$
[5]   **else if** $d[u] > rhs(u)$ **then**
[6] $d[u] := rhs(u)$
[7]   **fi**
[8]   **od**
end

Figure 6.2. An algorithm to update the unique fixed point of a collection of SWSF equations after a change in the collection of equations. Note that *rhs* and *key* are functions.

The correctness of the algorithm will follow from (1): it follows from the claim that line [6] can be executed at most once for each vertex u; line [4] too can be executed at most once for each vertex u, since once $d[u]$ is set to ∞, u cannot subsequently become an under-consistent vertex—as long as $d[u]$ is ∞, it cannot satisfy the condition in line [3]; hence $d[u]$ can only be changed in line [6], in which case the vertex becomes consistent and remains so, from claim (1). Hence, the algorithm makes at most two iterations for each vertex, and, hence, the algorithm must halt. The correctness follows immediately from the termination condition for the loop.

How can we establish that the algorithm is bounded? Claim (2) shows that the values of only affected variables are changed by the algorithm, and it follows from claim (1) that the algorithm makes at most two iterations for each affected variable. It follows that the algorithm makes a bounded number of iterations. We will later show that line [2] can be implemented to run in bounded time, which suffices to establish that the algorithm is bounded.

We will now prove claims (1) and (2). These claims follow from the fact that the keys of vertices chosen in line [2] over the iterations form a non-decreasing sequence. We first consider the change in the consistency status and the key value of vertices when a vertex u is processed (lines [3]-[7]) in some particular iteration. Let us denote the "initial" values of variables and expressions, that is, the value of these variables and expressions before the execution of lines [3]-[7] in the iteration under consideration, with the subscript "old", and to the "final" values of these variables and expressions with the subscript "new". In the following propositions u denotes the vertex chosen in line [2] of the particular iteration under consideration.

Proposition 6.1. If $rhs_{new}(w) \neq rhs_{old}(w)$ then $rhs_{new}(w) > key_{old}(u)$ and $rhs_{old}(w) > key_{old}(u)$.

Proof. Note that $rhs_{new}(w) = g_w(d[x_1], \ldots, d_{new}[u], \ldots, d[x_k])$, while $rhs_{old}(w) = g_w(d[x_1], \ldots, d_{old}[u], \ldots, d[x_k])$. It follows from Proposition 5.20(d)[2] that both $rhs_{old}(w)$ and $rhs_{new}(w)$ are greater than $min(d_{old}[u], d_{new}[u]) = min(d_{old}[u], rhs_{old}(u)) = key_{old}(u)$. $\square$

Proposition 6.2. If u is over-consistent at the beginning of the iteration, then it is consistent at the end of the iteration.

Proof. Since u is over-consistent initially, $d[u]$ will be assigned the value $rhs_{old}(u)$. As long as this assignment does not change the rhs value of u, u must be consistent. But $rhs_{old}(u)$ must be equal to $rhs_{new}(u)$—since, otherwise, we would have $rhs_{old}(u) > key_{old}(u) = rhs_{old}(u)$, which is a contradiction. $\square$

---

[2]This Proposition is reproduced below from the previous chapter:
Proposition 5.20(d): If g is a s.w.s.f. and $g(x_1, \ldots, x_i, \ldots, x_k) \neq g(x_1, \ldots, x_i', \ldots, x_k)$, then $g(x_1, \ldots, x_i, \ldots, x_k) > min(x_i, x_i')$ and, similarly, $g(x_1, \ldots, x_i', \ldots, x_k) > min(x_i, x_i')$.

Proposition 6.3. For any vertex w that is inconsistent at the end of the iteration, $key_{new}(w) \geq key_{old}(u)$.

Proof. Since, $key_{new}(w) = min(rhs_{new}(w), d_{new}[w])$, by definition, we need to show that (a) $rhs_{new}(w) \geq key_{old}(u)$, and that (b) $d_{new}[w] \geq key_{old}(u)$. Consider (a). If the rhs value of w did not change, then w must have been inconsistent originally. Hence, $rhs_{new}(w) = rhs_{old}(w) \geq key_{old}(w) \geq key_{old}(u)$. If the rhs value of w did change, then it follows from Proposition 6.1 that $rhs_{new}(w) \geq key_{old}(u)$. Now consider (b). If w was originally inconsistent, then $d_{new}[w] = d_{old}[w] \geq key_{old}(w) \geq key_{old}(u)$. If w was originally consistent, then $rhs(w)$ must have changed in value. It follows from Proposition 6.1 that $d_{new}[w] = d_{old}[w] = rhs_{old}(w) > key_{old}(u)$. $\square$

We now turn our attention to the change in the values and consistency statuses of variables change over the different iterations of the algorithm. The subscript i attached to any variable or expression denotes the value of the variable or expression at the beginning of iteration i.

Proposition 6.4. If $i < j$ then $key_i(u_i) \leq key_j(u_j)$. In other words, the keys of variables chosen in line [2] form a monotonically non-decreasing sequence.

Proof. This follows trivially from repeated applications of Proposition 6.3. $\square$

Proposition 6.5. Assume that the vertex u_i chosen in line [2] of the i-th iteration is an over-consistent vertex. Then, u_i remains consistent in all subsequent iterations. In particular, its value is never again changed.

Proof. We showed above in Proposition 6.2 that variable u_i is consistent at the end of the i-th iteration. It can never again become inconsistent because its rhs value can never again change—this follows because, if $rhs(u_i)$ were to change in a subsequent iteration, say the j-th iteration, then we would have $key_i(u_i) = rhs_i(u_i) = rhs_j(u_i) > key_j(u_j)$, from Proposition 6.1. But this contradicts Proposition 6.4. Since only the values of inconsistent variables are ever changed it follows that $d[u_i]$ is never again changed. $\square$

Proposition 6.6. Assume that the vertex u_i chosen in line [2] of the i-th iteration is an under-consistent vertex. Then, u_i is never assigned its original value, $d_i[u_i]$, again.

Proof. We need to show that the rhs value of variable u_i never becomes $d_i[u_i]$. This follows from Proposition 6.1 since if $rhs(u_i)$ changes in the j-th iteration, then we have $rhs_{j+1}(u_i) > key_j(u_j) \geq key_i(u_i) = d_i[u_i]$. $\square$

Proposition 6.7. Procedure *IncrementalFP* correctly computes the unique fixed point of the given collection of equations. Further, during the course of the computation it changes only the values of the variables that had an incorrect value at the beginning of the update. It also changes the value of a variable at most twice.

Proof. This proposition follows directly from Proposition 6.5 and Proposition 6.6, as explained earlier. $\square$

6.3. The Algorithm

In this section we present a detailed version of the algorithm we described in the previous section for the dynamic SWSF fixed point problem. The algorithm is described as procedure *DynamicSWSF–FP* in Figure 6.3. We assume that a dependence graph

---

procedure DynamicSWSF-FP (G, U)
declare
        G : a dependence graph of a set of SWSF equations
        U : the set of modified vertices in G
        u, v, w: vertices
        Heap: a heap of vertices
preconditions
        Every vertex in $V(G)-U$ is consistent
begin
[1]   Heap $:= \varnothing$
[2]   **for** $u \in U$ **do**
[3] $rhs[u] := g_u(d[x_1], \ldots, d[x_k])$
[4]   **if** $rhs[u] \neq d[u]$ **then**
[5]   InsertIntoHeap(*Heap, u*, min($rhs[u], d[u]$))
[6]   **fi**
[7]   **od**
[8]   **while** Heap $\neq \varnothing$ **do**
[9] $u :=$ ExtractAndDeleteMin(Heap)
[10]   **if** $rhs[u] < d[u]$ **then** /* u is overconsistent */
[11] $d[u] := rhs[u]$
[12]   **for** $v \in Succ(u)$ **do**
[13] $rhs[v] := g_v(d[x_1], \ldots, d[x_k])$
[14]   **if** $rhs[v] \neq d[v]$ **then**
[15]   AdjustHeap(*Heap, v*, min($rhs[v], d[v]$))
[16]   **else**
[17]   **if** $v \in$ *Heap* **then** Remove v from *Heap* **fi**
[18]   **fi**
[19]   **od**
[20]   **else** /* u is underconsistent */
[21] $d[u] := \infty$
[22]   **for** $v \in (Succ(u) \cup \{u\})$ **do**
[23] $rhs[v] := g_v(d[x_1], \ldots, d[x_k])$
[24]   **if** $rhs[v] \neq d[v]$ **then**
[25]   AdjustHeap(*Heap, v*, min($rhs[v], d[v]$))
[26]   **else**
[27]   **if** $v \in$ *Heap* **then** Remove v from *Heap* **fi**
[28]   **fi**
[29]   **od**
[30]   **fi**
[31]   **od**
end
postconditions
        Every vertex in $V(G)$ is consistent

Figure 6.3. An algorithm for the dynamic SWSF fixed point problem.

G of a collection of SWSF equations is given, and that every vertex u in the graph has a tentative output value $d[u]$. We assume that the set U of vertices whose associated equations have been modified is also part of the input to the algorithm. In other words, only vertices in U may be inconsistent. The other vertices are guaranteed to be consistent. This is the precondition for the algorithm to compute the correct solution to the modified set of equations.

The idea behind the algorithm was explained in the previous section. The algorithm maintains the following invariants, and the steps in the algorithm can be understood easier in terms of the invariants. The algorithm maintains a heap of all the inconsistent vertices—both over-consistent and under-consistent vertices—in the graph. An overconsistent vertex u occurs in the heap with a key (priority) of $g_u(d[x_1], \ldots, d[x_k])$, while an under-consistent vertex u occurs in the heap with a key value of $d[u]$. The heap is used to identify the inconsistency with the least key value at every stage. For every inconsistent vertex u, the algorithm also maintains $rhs[u]$, the value of the right-hand side of the equation associated with vertex u. Let us say a vertex u *satisfies the invariant* if (a) u occurs in *Heap* with key k iff u is an inconsistent vertex with $key(u) = k$, and (b) if u is an inconsistent vertex then $rhs[u] = g_u(d[x_1], \ldots, d[x_k])$.

Recall what each heap operation does. The operation *InsertIntoHeap* (H,i,k) inserts an item i into heap H with a key k. The operation *FindAndDeleteMin* (H) returns the item in heap H that has the minimum key and deletes it from the heap. The operation *AdjustHeap* (H,i,k) inserts an item i into *Heap* with key k if i is not in *Heap*, and changes the key of item i in *Heap* to k if i is in *Heap*.

We now verify that the algorithm does indeed maintain the invariants described above. Thus, we first need to show that all vertices satisfy the invariant whenever execution reaches line [8]. The precondition guarantees that all the initially inconsistent vertices must be in U. In lines [1]-[7], the algorithm creates a heap out of all the initially inconsistent vertices in the graph, and simultaneously the value $rhs[u]$ is properly defined for every inconsistent vertex u. Hence the invariant holds when execution reaches line [8] for the first time.

The loop in lines [8]-[31] processes and "fixes" the inconsistencies in the graph one by one, in increasing order of key value. An over-consistent vertex u is processed (lines [11]-[19]) by updating $d[u]$ to equal $g_u(d[x_1], \ldots, d[x_k])$, the value of the right-hand side of the equation associated with vertex u. This converts the over-estimated vertex u into a correct vertex. As a result of the assignment of a new value to $d[u]$ in line [11] some of the successors of u may fail to satisfy the invariant, though any vertex which is not a successor of u will continue to satisfy the invariant. When the loop in lines [12]-[19] completes execution all vertices are guaranteed to satisfy the invariant. In particular, lines [13]-[18] make sure v satisfies the invariant by computing its rhs value, determining its consistency status, and adjusting the heap.

An under-consistent vertex u is processed (lines [21]-[30]) by updating $d[u]$ to equal ∞, followed by an appropriate updating of the heap. This step converts an under-estimated vertex into either an over-estimated vertex or a correct vertex. Following the assignment of a new value to $d[u]$ in line [21], only u or some successor of u can fail to satisfy the invariant. These vertices are appropriately processed in lines [22]-[29], and hence the invariant is satisfied whenever execution reaches line [8].

To understand how the algorithm makes progress towards the correct solution consider how the correctness status of the vertices in the graph change. In each iteration of the loop in lines [8]-[31] the value, and hence the correctness status, of only one vertex (namely u) changes. In particular, in each iteration exactly one of the following happens. (1) An over-estimated vertex becomes correct. (2) An under-estimated vertex becomes over-estimated. (3) An under-estimated vertex becomes correct. In particular, the value of a correct vertex is never changed. An initially (*i.e.*, at the beginning of the algorithm) over-estimated vertex changes value exactly once. An initially under-estimated vertex changes values at most twice (either to the correct value ∞, or first to ∞ and then to the correct final value). It follows that the algorithm must terminate.

Since the heap is empty when the algorithm terminates, it follows immediately from the loop invariant that there exists no inconsistency in the graph when the algorithm halts. In particular, the computed d values form the unique fixed point of the collection Q of equations.

Let us now determine the time complexity of the algorithm. Let M_δ be a bound on the time required to compute the function associated with any vertex in CHANGED $\cup$ *Succ* (CHANGED). The initialization in lines [1]-[7] involves $|U|$ function evaluations and $|U|$ heap operations (insertions) and consequently takes $O(|U| \cdot (M_\delta + \log |U|))$ time, which is $O(|\delta| \cdot (M_\delta + \log |\delta|))$ time since U is MODIFIED$_\delta$.

Every vertex that is in the heap at some point during the execution must be an affected vertex or the successor of an affected vertex. Hence, the maximum number of elements in the heap at any point is $O(\|\delta\|)$, and every heap operation takes $O(\log \|\delta\|)$ time. It follows from the explanation given earlier that lines [11]-[19] are executed at most once for each affected vertex u. In these lines, the function associated with every vertex in *Succ* (u) is evaluated once, and at most $|Succ(u)|$ heap operations are performed. Hence, the lines [11]-[19] take $O(\|\{u\}\| \cdot (M_\delta + \log \|\delta\|))$ time (in one iteration). Lines [20]-[30] are similarly executed at most once for each affected vertex u. Consequently, lines [20]-[30] also take time $O(\|\{u\}\| \cdot (M_\delta + \log \|\delta\|))$ time (in one iteration).

Consequently, the whole algorithm runs in time $O(\|\delta\| \cdot (\log \|\delta\| + M_\delta))$, and the algorithm is a bounded scheduling cost algorithm.

6.4. An Improved Algorithm

The algorithm presented in the previous section is not the most efficient incremental algorithm for the SSSP>0 problem. The source of inefficiency is that the algorithm assumes that each function g_i is an *s.w.s.f.* and no more. The functions that arise in the shortest-path problem (and in any SSF grammar problem), however, have a special form. The function corresponding to a vertex u other than the source is $\min_{v \in Pred(u)} [d[v] + length(u \rightarrow v)]$. Such expressions permit the possibility of incre-

---

procedure DynamicSSF-G (G, P)
declare
       G : a SSF grammar;
       P : the set of modified productions in G
       GlobalHeap: a heap of non-terminals
       Heap: array[Nonterminals] of heap of productions;
       SP: array[Nonterminals] of set of productions
preconditions: Every production in $G–P$ is consistent. (See Definition 6.8)

       procedure recomputeProductionValue(p : a production)
       begin
[1]   let p be the production $Y \longrightarrow g(X_1, \ldots, X_k)$
[2] $value = g(d[X_1], \ldots, d[X_k])$
[3]   **if** ($value < d[Y]$) **then**
[4]   AdjustHeap($Heap[Y], p, value$)
[5]   **else**
[6]   **if** $p \in Heap[Y]$ **then** Remove p from $Heap[Y]$ **fi**
[7]   **fi**
[8]   **if** ($value \le d[Y]$) **then** $SP[Y] := SP[Y] \cup \{p\}$ **else** $SP[Y] := SP[Y]-\{p\}$ **fi**
[9]   **if** ($SP[Y] = \emptyset$) **then**   /* Y is under-consistent */
[10]   AdjustHeap($GlobalHeap, Y, d[Y]$)
[11]   **elseif** $Heap[Y] \ne \emptyset$ **then**   /* Y is over-consistent */
[12]   AdjustHeap($GlobalHeap, Y, min-key(Heap[Y])$)
[13]   **else** /* Y is consistent */
[14]   **if** $Y \in GlobalHeap$ **then** Remove Y from $GlobalHeap$ **fi**
[15]   **fi**
       end

begin
[16]   GlobalHeap := $\emptyset$
[17]   **for** every production $p \in P$ **do**
[18]   recomputeProductionValue(p)
[19]   **od**
[20]   **while** GlobalHeap $\ne \emptyset$ **do**
[21]   Select and remove from GlobalHeap a non-terminal X with minimum key value
[22]   **if** $key(X) < d[X]$ **then** /* X is overconsistent */
[23] $d[X] := key(X)$
[24] $SP[X] := \{p \mid p$ is a production for X such that $value(p) = d[X]\}$
[25] $Heap[X] := \emptyset$
[26]   **for** every production p with X on the r.h.s. **do** recomputeProductionValue(p) **od**
[27]   **else** /* X is underconsistent */
[28] $d[X] := \infty$
[29] $SP[X] := \{p \mid p$ is a production for $X\}$
[30] $Heap[X] := makeHeap(\{p \mid p$ is a production for X with $value(p) < d[X]\})$
[31]   **if** $Heap[X] \ne \emptyset$ **then** AdjustHeap($GlobalHeap, X, min-key(Heap[X])$) **fi**
[32]   **for** every production p with X on the r.h.s. **do** recomputeProductionValue(p) **od**
[33]   **fi**
[34]   **od**
end
postconditions: Every non-terminal and production in G is consistent

---

Figure 6.4. An algorithm for the dynamic SSF grammar problem.

mental computation of the expression itself. For instance, evaluating this value from scratch takes time $\Theta(|Pred(u)|)$, while if the value of this expression is known, and the value of $d[v]$ decreases for some $v \in Pred(u)$, the new value of the expression can be recomputed incrementally in constant time. Note that this kind of incremental recomputation of an expression's value is performed repeatedly in Dijkstra's algorithm for the batch SSSP$\geq$0 problem. Unfortunately, an incremental algorithm for the SSSP problem has to also contend with the possibility that the value of $d[v]$ *increases* for some $v \in Pred(u)$. The need to maintain the value of the expression $\min\limits_{v \in Pred(u)} [d[v] + length(u \longrightarrow v)]$ as the values of $d[v]$ change immediately suggests the possibility of maintaining the set of all values $\{ d[v] + length(u \longrightarrow v) \mid v \in Pred(u) \}$ as a heap. Our approach is to maintain a particular subset of the set $\{ d[v] + length(u \longrightarrow v) \mid v \in Pred(u) \}$ as a heap, since maintaining the whole set as a heap requires unnecesary work.

In this section we present a more efficient version of algorithm *DynamicSWSF–FP* that utilizes the special form of the equations induced by the SSF grammar problem. The algorithm is described as procedure *DynamicSSF–G* in Figure 6.4. The algorithm, as presented, addresses the dynamic SSF grammar problem, and, hence, might appear to be less general than the algorithm presented in the previous section, which addresses the dynamic SWSF fixed point problem. Procedure *DynamicSSF–G* can, in fact, be used for the dynamic SWSF fixed point problem with some simple modifications, though it will be more efficient than procedure *DynamicSWSF–FP* only when the equations have the special form described above. We address the less general SSF grammar problem here since it is this problem that motivates the improvements to the algorithm, but emphasize that the improved algorithm is as general as the original algorithm in terms of the class of problem instances that it can handle. For this reason we refer to procedure *DynamicSSF–G* as an improvement of *DynamicSWSF–FP* rather than merely a specialization of *DynamicSWSF–FP*.

We first explain the idea behind the algorithm, then prove the correctness of the algorithm, and finally analyze its time complexity.

We assume that an SSF grammar is given, and that every non-terminal X in the grammar has a tentative output value $d[X]$. We assume that the change in the input takes the form of a change in some of the productions and production functions of the grammar. This type of modification is general enough to include insertions and deletions of productions as well, since a non-existent production can be treated as a production whose production function is the constant-valued function ∞. The insertion or deletion of non-terminals can be handled just as easily. So we assume that the input to the algorithm includes a set P of productions whose production functions have been modified.

The steps given in lines [16]-[34] implement essentially the same idea as procedure *DynamicSWSF–FP*. A heap, called *GlobalHeap*, of all the inconsistent non-terminals is maintained as before, and in each iteration the inconsistent non-terminal X with the least key is processed, just as before. In *DynamicSWSF–FP* a change in

the value of a vertex is followed by the complete re-evaluation of the function associated with the successors of that vertex, in order to identify the change in the consistency status of those vertices. This is the step that the new algorithm, procedure
DynamicSSF–G, performs differently. The new algorithm identifies changes in the
consistency status of other non-terminals in an *incremental* fashion. We now
describe the auxiliary data structures that the algorithm uses to do this. These auxiliary data structures are retained across invocations of the procedure.

Note that the value associated with a non-terminal X is $d[X]$. We define the
value of a production $Y \rightarrow g(X_1, \ldots, X_k)$ to be $g(d[X_1], \ldots, d[X_k])$. For every
non-terminal X, the algorithm maintains a set $SP[X]$ of all productions with X as the
left-hand side whose value is less than or equal to $d[X]$. The algorithm also maintains for every non-terminal X a heap *Heap*$[X]$ of all the productions with X as the
left-hand side whose value is strictly less than $d[X]$, with the value of the production
being its key in the heap.

Consider a production $p = Y \rightarrow g(X_1, \ldots, X_k)$. We say that the production p
satisfies the invariant if (a) $p \in SP[Y]$ iff *value*$(p) \le d[Y]$ and (b) $p \in Heap[Y]$ iff
value$(p) < d[Y]$. Thus, we want to maintain $SP[Y]$ and *Heap*$[Y]$ such that all productions satisfy the invariant. However, both at the beginning of the update and temporarily during the update, several productions may fail to satisfy the invariant.

We use these auxiliary data structures to determine the consistency status of
non-terminals. Note that a non-terminal X is under-consistent iff $SP[X]$ is empty and
$d[X] < \infty$,[3] in which case its key is $d[X]$; X is over-consistent iff *Heap*$[X]$ is non-
empty, in which case its key is given by $min–key(Heap[X])$, the key of the item with
the minimum key value in *Heap*$[X]$. The invariant that *GlobalHeap* satisfies is that
every non-terminal X for which $SP[X]$ is empty and $d[X]$ is less than ∞ occurs in
GlobalHeap with a key of $d[X]$, while every non-terminal X for which *Heap*$[X]$ is
non-empty occurs in *GlobalHeap* with a key of $min–key(Heap[X])$. It follows from
the preceeding explanation that *GlobalHeap* consists of exactly the inconsistent non-
terminals with their appropriate keys.

We now show that the algorithm maintains these data structures correctly and
that it updates the solution correctly. However, we first need to understand the
precondition these data structures will have to satisfy at the beginning of the algorithm.

Definition 6.8. A production $p = Y \rightarrow g(X_1, \ldots, X_k)$ is said to be *consistent* if (a)
$p \notin Heap[Y]$ and (b) either *value*$(p) = d[Y]$ and $p \in SP[Y]$ or *value*$(p) > d[Y]$
and $p \notin SP[Y]$. In other words, p is consistent iff it satisfies the invariant and, in
addition, *value*$(p) \ge d[Y]$.

The precondition we assume to hold at the beginning of the update is that
every unmodified production is consistent. The invariant the algorithm maintains is

---

[3] In general, the condition that $SP[X]$ be empty subsumes the condition that $d[X]$ be less than
∞. The latter condition is relevant only if no production has X on the left-hand side.

that whenever execution reaches line [20] every production satisfies the invariant, and that the *GlobalHeap* contains exactly the inconsistent non-terminals. The postcondition established by the algorithm is that every production and non-terminal in the grammar will be consistent.

The procedure *recomputeProductionValue(p)* makes production p consistent by computing its value (in line [2]) and updating the data structures $SP[Y]$ (line [8]) and *Heap*$[Y]$ (lines [3]-[7]) appropriately, where Y is the left-hand side of p. These changes are followed by appropriate updates to *GlobalHeap* in lines [9]-[15].

We now show that whenever execution reaches line [20] every production satisfies the invariant, and *GlobalHeap* contains exactly the inconsistent non-terminals. The lines [16]-[19] initially establish the invariant. Subsequently, in each iteration of the loop in lines [20]-[38], whenever the value of a non-terminal changes (either in line [23] or line [30]) procedure *recomputeProductionValue(p)* is called for every production p that might have become inconsistent. Thus, the invariant is re-established.

It follows from the explanation in the previous paragraph that every non-terminal and production in the grammar is consistent when the algorithm halts.

Let us now consider the time complexity of the improved algorithm. In Algorithm *DynamicSWSF–FP* the individual equations were treated as indivisible units, the smallest units of the input that could be modified. The algorithm outlined in this section, however, specifically deals with the equations generated by an SSF grammar. A finer granularity of input modifications is made possible by allowing individual productions to be modified. Consequently, it is necessary to consider a refined version of the dependence graph in analyzing the time complexity of the algorithm.

The bipartite graph $B = (N, P, E)$ consists of two disjoint sets of vertices N and P, and a set of edges E between N and P. The set N consists of a vertex n_X for every non-terminal X in the grammar, while the set P consists of a vertex n_p for every production p in the grammar. For every production p in the grammar, the graph contains an edge $n_X \rightarrow n_p$ for every non-terminal X that occurs on the right-hand side of p, and an edge $n_p \rightarrow n_Y$ where Y is the left-hand side non-terminal of p. The set AFFECTED consists of the set of all vertices n_X where X is a non-terminal whose output value changes, while the set MODIFIED consists of the set of all vertices n_p, where p is a modified production. The set CHANGED is AFFECTED $\cup$ MODIFIED.

Let us first consider the time spent in the main procedure, namely lines [16]-[38]. As explained in the previous section, the loop in lines [20]-[38] iterates at most $2 \cdot |\text{AFFECTED}|$ times. Lines [23]-[28] are executed at most once for every affected non-terminal X, while lines [30]-[36] are similarly executed at most once for every affected non-terminal X. Consequently, the steps executed by the main procedure can be divided into (a) $O(\|\text{CHANGED}\|_B)$ invocations of the procedure *recomputeProductionValue* (lines [18], [27] and [35]), (b) $O(|\text{AFFECTED}|)$ operations on *GlobalHeap* (line [21]), and (c) the remaining steps, which take time $O(\|\text{CHANGED}\|_B)$.

Let us now consider the time taken by a single execution of procedure *recomputeProductionValue*. The procedure essentially performs (a) one function computa-

tion (line [2]), (b) $O(1)$ set operations (lines [8] and [9]), (c) $O(1)$ $Heap[Y]$ operations (lines [4] or [6]), and (d) $O(1)$ $GlobalHeap$ operations (lines [10], [12] or [14]).
The set operations on $SP[Y]$ can be done in constant time by associating every production $Y \rightarrow g(X_1, \ldots, X_k)$ with a bit that indicates if it is in the set $SP[Y]$ or not. It
can be easily verified that each $Heap[Y]$ and $GlobalHeap$ have at most
$\|\text{AFFECTED}\|_B$ elements. Consequently, each heap operation takes at most
$\log \|\text{AFFECTED}\|_B$ time.

As before, let $M_{B,\delta}$ be a bound on the time required to compute the production
function associated with any production in $\text{CHANGED} \cup Succ(\text{CHANGED})$. Then,
procedure $recomputeProductionValue$ itself takes time $O(\log \|\delta\|_B + M_{B,\delta})$. Hence,
the whole algorithm runs in time $O(\|\delta\|_B \cdot (\log\|\delta\|_B + M_{B,\delta}))$.

Let us now consider the SSSP>0 problem. Each production function can be
evaluated in constant time in this case, and, hence, the algorithm runs in time
$O(\|\delta\| \log\|\delta\|)$. (Note that in the case of the SSSP>0 problem the input graph G
and the bipartite graph B are closely related, since each "production" vertex in B
corresponds to an edge in G. Hence, $\|\delta\|_B = O(\|\delta\|_G)$.)

We now consider a special type of input modification for the SSSP>0 problem
for which it is possible to give a better bound on the time taken by the update algorithm. Assume that the change in the input is a homogeneous decrease in the length
of one or more edges. In other words, no edges are deleted and no edge-length is
increased. In this case it can be seen that no under-consistent vertex exists, and that
the value of no vertex increases during the update. In particular, the $AdjustHeap$
operations (in lines [4], [10], and [12]) either perform an insertion or decrease the key
of an item. Lines [6] and [14] are never executed. Consequently, procedure $recom$-
$puteProductionValue$ takes time $O(1)$ if relaxed heaps [Dri88] or Fibonacci heaps
[Fre87] are used. (In the latter case, the time complexity is the amortized complex-
ity.) It can also be verified that the number of elements in any of the heaps is $O(|\delta|)$.
Hence, the algorithm runs in time $O(\|\delta\| + |\delta| \log|\delta|)$. In particular, if m edges are
inserted into an empty graph with n vertices, the algorithm works exactly like the
$O(m + n \log n)$ implementation of Dijkstra's algorithm due to Fredman and Tarjan
[Fre87]. The asymptotic complexity of the algorithm can be further improved by
using the recently developed AF-heap data structure [Fre90a].

6.5. Extensions to the Algorithm
In this section, we briefly outline various possible extensions and applications of the
incremental algorithms described in the previous sections.

6.5.1. Answering Queries on Demand
The incremental algorithms presented in this chapter update the solution to the whole
problem instance when they are invoked. This can potentially result in unnecessary
computation being performed, since (potentially large) parts of the computed solution
may never be used before they are "affected" by subsequent modifications to the
input. In such situations it may be appropriate to use a demand-driven algorithm,
where the solution to the problem instance is computed as and when necessary.

Procedures *DynamicSSF–G* and *DynamicSWSF–FP* can be easily adapted to work in such a demand-driven fashion. Consider the update-query model of dynamic algorithms commonly used: assume that the algorithm has to process a sequence of requests, where each request is either an update to the input grammer (set of equations, in the case of a fixed point problem) or a query asking for the cost of the optimal derivation from a specific non-terminal (value of a variable). For each update operation, the algorithm does nothing more than "note" down the actual modification performed. When a query is performed, the algorithm starts processing the sequence of updates performed since the last time a query was processed by invoking *DynamicSSF–G* (*DynamicSWSF–FP*). However, instead of running the algorithm to completion, we may stop the processing when we know the correct answer to the specific query being processed. It follows from the earlier discussion (see, in particular, Proposition 6.1 and the proof of Proposition 6.5) that any consistent vertex whose value is less than $key(u)$, where u is the inconsistent vertex with the minimum key value, is, in fact, correct. Consequently, in this version of the dynamic SSF grammar problem, for example, if the query asks for the value $d[Y]$ of some non-terminal Y, then the updating algorithm in Figure 6.4 may stop when the key of the non-terminal X selected in line [21] is greater than $d[Y]$.

Some other minor modifications to the algorithms are necessary. For example, in Figure 6.4, *GlobalHeap*, the heap of inconsistent vertices, need no longer be empty in between updates or queries. Consequently, the algorithm should no longer initialize *GlobalHeap* to be empty at the beginning of the update, but just carry the value over from the previous invocation of the algorithm.

6.5.2. Maintaining Minimum Cost Derivations

We have so far considered only the problem of maintaining the *cost* of the minimum cost derivations, and not the problem of maintaining minimum cost *derivations* themselves. However, the algorithm outlined in the previous section can be easily extended to maintain the minimum cost derivations too. The set $SP[X]$ computed by the algorithm is the set of all productions for X that can be utilized as the first production in minimum cost derivations of terminal strings from X. Hence, all possible minimum cost derivations from a non-terminal can be recovered from this information. In particular, consider the SSSP>0 problem. Every production p for a non-terminal N_v corresponds to an incoming edge $u \longrightarrow v$ of vertex v, where v is a vertex other than the source. The production p will be in $SP[N_v]$ iff a shortest path from the source to u followed by the edge $u \longrightarrow v$ yields a shortest path from the source to v. Hence, a single shortest-path from the source vertex to any given vertex can be identified in time proportional to the number of edges in that path, provided the set $SP[X]$ is implemented as a doubly linked list so that an arbitrary element from the set can be chosen in constant time.

6.5.3. The All-Pairs Shortest-Path Problem

We have seen that the algorithm outlined in the previous section can be used to update the solution to the single-source (or the single-sink) shortest-path problem when the underlying graph undergoes modifications. We briefly sketch how this algorithm can be adapted to update the solution to the all-pairs shortest-path problem too. The essential approach is to make repeated use of our incremental algorithm for the SSSP>0 problem. However, it is not necessary to update the single-source solution for every vertex in the graph; it is possible to identify a subset of the vertices for which it is sufficient to update the single-source solution. Let $u_i \rightarrow v_i$, for $1 \leq i \leq k$, be the set of modified (inserted or deleted) edges. Let $d(x,y)$ denote the length of a shortest path from x to y. Then, for any two vertices s and t, $d(s,t)$ can change only if for some $i \in [1,k]$ both $d(s,v_i)$ and $d(u_i,t)$ change. Hence, by updating the single-source solution for every u_i, we can identify the set of vertices t for which the single-sink solution will change. Similarly, by updating the single-sink solution for every v_i, we can identify the set of vertices s for which the single-source solution will change. Then, we can update the single-sink solution and the single-source solution only for those vertices for which the solution can change. However, we note that for certain special cases, such as updating the solution to the APSP>0 problem after the insertion of an edge, this approach does not yield the best possible incremental algorithm. The algorithm presented in Chapter 4 works better in this special case.

6.5.4. Handling Edges of Non-Positive Length

Consider the use of our incremental algorithm for the SSSP problem. The proof of correctness of our algorithm and the analysis of its time complexity both rely on the fact that all edges have a positive length. We now discuss some types of input changes for which this restriction on the edge lengths can be somewhat relaxed. We first consider zero-length edges. It can be shown that if the change in the input graph is a homogeneous decrease in the length of one or more edges then the algorithm works correctly as long as all edges have a non-negative length (*i.e.*, zero-length edges do not pose a problem). Similarly, if the input change is a homogeneous increase in the length of one or more edges then the algorithm works correctly as long as all edges have a non-negative length and there are no cycles in the graph of zero length (*i.e.*, zero-length edges do not pose a problem as long as no zero-length cycles exist in the graph).

We now consider negative length edges. For certain types of input modifications it is possible to use a variant of our incremental algorithm to update the solution to the SSSP problem (with arbitrary edge lengths), as long as all cycles in the graph have a positive length. The idea, which was discussed earlier in Chapter 4, is to adapt the technique of Edmonds and Karp for transforming the length of every edge to a non-negative real without changing the graph's shortest paths [Edm72, Tar83]. Their technique is based on the observation that if f is any function that maps vertices of the graph to reals, and the length of each edge $a \rightarrow b$ is replaced by $f(a) + length(a \rightarrow b) - f(b)$, then the shortest paths in the graph are unchanged from the original edge-length mapping. If f satisfies the property that $f(a) + length(a \rightarrow b) - f(b) \geq 0$ for every edge $a \rightarrow b$ in the graph, then the transformed length of every edge will be positive.

Now consider the incremental SSSP problem. Let $d_{old}(u)$ denote the length of the shortest path in the input graph G from $source(G)$ to u before G was modified. Consider the effect of the above edge-length transformation if we simply define $f(u)$ to be $d_{old}(u)$. First note that the transformation is well-defined only for edges $a \rightarrow b$ such that $d_{old}(b)$ is not ∞. For every edge $a \rightarrow b$ in the original graph we have $d_{old}(b) \leq d_{old}(a) + length_{old}(a \rightarrow b)$. Consequently, $d_{old}(a) + length_{old}(a \rightarrow b) - d_{old}(b) \geq 0$. Hence, the transformed length of an edge $a \rightarrow b$ will be non-negative as long as $length_{new}(a \rightarrow b) \geq length_{old}(a \rightarrow b)$ (i.e., as long as the length of the edge $a \rightarrow b$ was not decreased during the input modification), and $d_{old}(b)$ is not ∞.

In particular, this idea can be used to adapt our incremental algorithm to update the solution to the SSSP problem when the lengths of a collection of edges are increased (possibly to ∞), and no edge is inserted or no edge-length is decreased. This will work since the length of an edge $a \rightarrow b$ is relevant only if a can be reached from the source vertex and, hence, only if both $d_{old}(a)$ and $d_{old}(b)$ are finite. The transformed length of all such edges are non-negative, and our incremental algorithm is applicable as long as there are no cycles of zero length in the graph. Note that it is not necessary to compute the transformed length for all edges at the beginning; instead, the transformed length of an edge can be computed as and when the length of that edge is needed. This is essential to keep the algorithm a bounded one.

The technique of edge-length transformation can also be used in a special case of edge insertion or edge-length decrease. Assume that the length of a set of edges F, all directed to a specific vertex u that was already reachable from the source, are decreased (possibly from ∞). The above edge-length transformation makes the lengths of all *relevant* edges non-negative. The transformed length of the edges in F are not guaranteed to be non-negative; however, this causes no difficulties because edges directed to u are in a sense irrelevant to the updating algorithm. More details can be found in Chapter 4.

6.6. Related Work

In this chapter we have presented an incremental algorithm for the dynamic SWSF fixed point problem. The dynamic SWSF fixed point problem includes the dynamic SSF grammar problem as a special case, which, in turn, includes the dynamic SSSP>0 problem as a special case. Thus, we obtain an incremental algorithm for the dynamic SSSP>0 problem as a special case of algorithm *DynamicSSF–G*, which was described in Section 5. We have also described how the algorithm can be generalized to handle negative edge lengths under certain conditions, and how the algorithm for the dynamic single-source shortest-path problem can be utilized for the dynamic all-pairs shortest-path problem as well.

Knuth [Knu77] introduced the grammar problem as a generalization of the shortest-path problem, and generalized Dijkstra's algorithm to solve the batch SF grammar problem. We know of no previous work on incremental algorithms for the dynamic grammar problem.

Recently, Ausiello *et al.* [Aus92] presented a *semi-dynamic* algorithm for maintaining optimal hyperpaths in directed hypergraphs. This algorithm is quite similar to our algorithm, except that it handles only the insertion of hyperarcs (productions) into the hypergraph (grammar). The relationship between the hypergraph problem and grammar problem is discussed below.

A directed hypergraph consists of a set of nodes and a set of hyperarcs. Each hyperarc connects a set of sources nodes to a single target node. The concept of hyperpaths is defined recursively. There exists an empty hyperpath from a set S of nodes to a node t if $t \in S$. A non-empty hyperpath from a set S of nodes to a node t consists of an hyperarc from a set S' to t and a hyperpath from S to s for every node s in S'. Ausiello *et al.* [Aus92] introduced the concept of a *value-based measure* for hyperpaths. Assume that every hyperarc e has an associated "weight" $wt(e)$. A value-based measure μ is described by a triple (f, ψ, μ_0), where μ_0 is a real value, f is a monotonic, binary, real-valued function, and ψ is a monotonic, commutative, and associative function from sets of reals to reals. The measure $\mu(\varnothing)$ of an empty hyperpath $\varnothing$ is defined to be μ_0; the measure $\mu(P)$ of a non-empty path P that can be recursively decomposed into a hyperedge e and hyperpaths $P_1, \ldots, P_k$ is defined to be $f(wt(e), \psi(\mu(P_1), \ldots, \mu(P_k)))$.

The analogy between context-free grammars and directed hypergraphs should be obvious. Nodes correspond to non-terminals, while hyperarcs correspond to productions. A *hyperpath* in a hypergraph corresponds to a *derivation* in the grammar. The value-based measure of a hyperpath is similar to the cost assigned to a derivation in Knuth's grammar problem, and an optimal hyperpath corresponds to a minimum-cost derivation.

The grammar problem is actually a strict generalization of the (unordered) directed hypergraph problem. In particular, in the grammar problem a richer class of functions are permitted on the hyperedges (and hence as "value-based measure functions"). This comes about because of two ways in which the frameworks differ:

(i)   In the grammar problem, different productions θ and τ can have different production functions g_θ and g_τ. In the (unordered) directed hypergraph problem, all hyperedges have functions built from a single function f and a single function ψ.

(ii)   In the grammar problem the nonterminals of each production —*i.e.*, the predecessors in each hyperedge—are ordered. Because the predecessors in each hyperedge are unordered, the function ψ, is required to be commutative and associative. This is not necessary in the grammar problem because the order of nonterminals in productions permits making some distinctions among the values that "flow" along the hyperedges. Thus, for example, with the grammar-problem formulation one can use a function such as:
$$g_\theta \triangleq \lambda x, y \cdot w_\theta + 2x + y.$$
This is not possible in the class of unordered problems that Ausiello et al. deal with.

There is another minor difference between the grammar problem and the hypergraph problem worth mentioning. Derivations of terminal strings from non-terminals really correspond to hyperpaths from an empty set of nodes to a node. Hyperpaths from an arbitrary, non-empty, set of nodes to a node can model "partial derivations", or derivations of sentences containing non-terminals as well. This is, however, not a significant difference. Such hyperpaths can be represented in the grammar by derivations by simply adding epsilon productions for all the relevant non-terminals.

Previous work on algorithms for the dynamic shortest-path problem[4] includes papers by Murchland [Mur, Mur67], Loubal [Lou67], Rodionov [Rod68], Halder [Hal70], Pape [Pap74], Hsieh *et al.* [Hsi76], Cheston [Che76], Dionne [Dio78], Goto *et al.* [Got78], Cheston and Corneil [Che82], Rohnert [Roh85], Even and Gazit [Eve85], Lin and Chang [Lin90], Ausiello *et al.* [Aus90, Aus91], and Ramalingam and Reps [Ram91] (see Chapter 4). These algorithms may be classified into groups based on (a) the information computed by the algorithm (such as the whether the all-pairs or single-source version of the problem is addressed), (b) the assumptions made about the edge lengths, and (c) the type of modification that the algorithm handles. What distinguishes the work reported in this chapter from all of the work cited above is that it is the first incremental algorithm that places no restrictions on how the underlying graph can be modified between updates. Our work addresses the single-source shortest-path problem with the restriction that all edges be positive in detail. It also briefly addresses the dynamic all-pairs shortest-path problem and extensions to handle edges with non-positive length.

The remainder of this section provides a brief overview of the different groups of dynamic shortest-path algorithms, the different techniques used by the various algorithms, and a brief comparison of the different algorithms. The table in Figure 6.5 summarizes this discussion. We remind the reader that our comments about the cases of an edge-insertion or an edge-deletion apply equally well to the cases of a decrease in an edge length and an increase in an edge length, respectively.

We begin with the version of the problem that has been studied the most, namely the all-pairs version. Given a graph G and a modification δ to the graph, let $d_{old}(x,y)$ and $d_{new}(x,y)$ denote the length of a shortest path from x to y in the graphs G and $G + \delta$ respectively. The pair (x,y) is said to be an *affected pair* if $d_{new}(x,y)$ is different from $d_{old}(x,y)$. A vertex x is said to be an *affected source* if there exists a vertex y such that (x,y) is an affected pair; similarly x is said to be an *affected sink* if there exists a vertex y such that (y,x) is an affected pair.

Let us now consider the problem of processing the insertion of an edge $u \rightarrow v$. This problem is in some sense the easiest among the various versions of the dynamic shortest-path problem; at least, it is fairly straight-forward to determine

---

[4]More recently, Frigioni *et al.* [Fri94] and Djidjev *et al.* [Dji95] have also presented algorithms for some versions of the shortest path problem.

Problem	Modifications	Best bounded algorithm(s)	Other bounded algorithms	Unbound-ed algorithms
APSP	Single Edge Insertion	[Lin90], [Aus91]	Chapter 4, [Roh85], [Eve85]	[Dio78], [Rod68], [Lou67], [Mur67]
APSP	Single Edge Deletion			[Roh85], [Eve85], [Dio78], [Rod68], [Mur67]
APSP-Cycle>0	Single Edge Deletion	Chapter 4		
APSP>0	Single Edge Deletion			[Hal70]
APSP>0	Arbitrary Modification	This chapter		
Multiple SSSP	Multiple Edge Insertions			[Got78]
SSSP>0	Arbitrary Modification	This chapter		
SSSP-Cycle>0	Multiple Edge Deletions	This chapter		
SSSP	Restricted Edge Insertion	This chapter		[Got78]

Figure 6.5. Various versions of the dynamic shortest-path problem and incremental algorithms for them. Note that APSP>0 and SSSP>0 refer to problems where every edge is assumed to have positive length, while APSP-Cycle>0 and SSSP-Cycle>0 refer to problems where every cycle is assumed to have positive length, with no restrictions on edge lengths. The modification referred to in the last item of the table, namely "restricted edge insertion", is the insertion of one or more edges, all directed to the same vertex, a vertex that must already be reachable from the source.

$d_{new}(x,y)$ in constant time, for any given pair of vertices (x,y) since
$$d_{new}(x,y) = \min (\, d_{old}(x,y),\, d_{old}(x,u) + length_{new}(u \longrightarrow v) + d_{old}(v,y)\,).$$
Computing $d_{new}(x,y)$ for every pair of vertices (x,y) using the above equation takes time $O(n^2)$, which is better than the time complexity of the best batch algorithm for APSP. Most of the known algorithms for this problem do even better by first identifying an approximation A to the set of all affected pairs and then updating $d(x,y)$ only for $(x,y) \in A$. The best algorithm currently known for this problem, developed independently by Lin and Chang [Lin90] and Ausiello et al. [Aus91], restricts the set of pairs of vertices for which the d value is recomputed by a careful traversal of the shortest-path trees of the graph before the modification. The algorithm due to Even and Gazit [Eve85] is similar and identifies the same set of pairs of vertices but is slightly less efficient since it does not maintain shortest-path trees. The algorithms presented in Rohnert [Roh85] and Chapter 4 are based also on similar ideas. All of

the above algorithms are bounded algorithms. It is worth mentioning that the improved efficiency of the algorithms described in [Lin90] and [Aus91] is obtained at a cost: these algorithms make use of the shortest-path-tree data structure, the maintenance of which can make the processing of an *edge-deletion* more expensive. The algorithms due to Murchland [Mur67], Dionne [Dio78], and Cheston [Che76] are all based on the observation that x is an affected source [sink] iff (x,v) $[(u,x)]$ is an affected pair. These algorithms identify the set of affected sources S_1 and the set of affected sinks S_2 in $O(n)$ time using equation (‡), and use $S_1 \times S_2$ as an approximation to the set of affected pairs. Consequently, these algorithms are unbounded.

Let us now consider the problem of processing the deletion of an edge $u \rightarrow v$ from the graph. Edge deletion is not as easy to handle as edge insertion. As Spira and Pan [Spi75] show, the batch all-pairs shortest-path problem can, in some sense, be reduced to the problem of updating the solution to the all-pairs shortest-path problem after an edge deletion. An incremental algorithm that saves only shortest-path information cannot, in the worst case, do any better than a batch algorithm, which is not true in the case of edge insertion.

Most algorithms for processing an edge deletion follow the approach of first identifying an approximation A to the set of all affected pairs, and then computing the new d value for every affected pair. We showed in Chapter 4 show that it is possible to identify the set of affected pairs exactly if the graph does not have zero-length cycles, and described the only known bounded incremental algorithm for this problem. This algorithm is based on the repeated application of a bounded algorithm for the dynamic SSSP>0 problem (see below). The set of all affected sinks is identified by using the algorithm for the dynamic SSSP>0 problem with u as the source, since x is an affected sink iff (u,x) is an affected pair. The APSP solution can then be updated by updating the single-sink solution for every affected sink.

The algorithms due to Rohnert [Roh85] and Even and Gazit [Eve85] can also be viewed as consisting of the repeated application of an algorithm for the dynamic SSSP problem, though they are not described as such. These algorithms, however, do not identify the set of affected pairs exactly. A vertex pair (x,y) is treated as a possibly affected pair iff $u \rightarrow v$ is in the current shortest path from x to y that the algorithm maintains. (Note that an alternative shortest path from x to y that does not contain edge $u \rightarrow v$ might exist in the original graph, and hence (x,y) might not be an affected pair.) However, these algorithms have the advantage that they work even in the presence of zero-length cycles.

All the above-mentioned algorithms use an adaptation of Dijkstra's algorithm to solve the dynamic SSSP algorithm. The algorithms can, however, be adapted to handle negative length edges using the technique outlined in Section 6.5.3. The algorithms due to Rodionov [Rod68], Murchland[Mur67], Dionne [Dio78], and Cheston [Che76], are all based on a different, and less efficient, technique of computing the new d value for every pair in A, the approximation to the set of affected pairs, using an adaptation of Floyd's algorithm for the batch shortest-path problem. A vertex pair (x,y) is considered to be possibly affected and is included in A iff $d_{old}(x,y) = d_{old}(x,u)$ + $length_{old}(u \rightarrow v)$ + $d_{old}(v,y)$. The adapted version of Floyd's algorithm differs from the original version in that in each of the n iterations only the d values of vertex pairs in A are recomputed. This algorithm runs in $O(|A| \cdot n)$ time.

Let us now consider the problem of updating the solution to the APSP problem after non-unit changes to the graph. This problem has not received much attention. The algorithm outlined in Section 6.5.2 for the dynamic APSP>0 problem is the only known incremental algorithm for any version of the dynamic APSP problem that is capable of handling insertions and deletions of edges simultaneously. Goto and Sangiovanni-Vincentelli [Got78] outline an incremental algorithm for updating the solution to multiple SSSP problems on the same graph—that is, the shortest-path information for a given set of sinks—when the lengths of one or more edges in the graph are decreased. Rodionov [Rod68] considers the problem of updating the solution to the APSP problem when the lengths of one or more edges all of which have a common endpoint are decreased.

Versions of the shortest-path problem other than the all-pairs version have not received much attention either. Goto and Sangiovanni-Vincentelli [Got78] consider the dynamic version of the problem of solving multiple single-source shortest-path problem: given a graph G and a *set* of source vertices S, determine the length of the shortest path between s and u for every source vertex s and every vertex u. Hence, the algorithm in [Got78] applies to the single-source problem as a special case.

Loubal [Lou67], and Halder [Hal70] study a generalization of the all-pairs shortest-path problem, where a subset S of the vertices in the graph is specified and the shortest path between any two vertices in S have to be computed.

In conclusion, the work described in this chapter differs from the previous work in this area in several ways. First, the incremental algorithm we have presented is first algorithm for any version of the dynamic shortest-path problem that is capable of handling arbitrary modifications to the graph (*i.e.*, multiple heterogeneous changes to the graph). Second, the version of the dynamic shortest-path problem we address, namely the single-source version, has been previously considered only in [Got78]. The algorithm described in this chapter is more efficient and capable of handling more general modifications than the algorithm described in [Got78]. (However, the latter algorithm, unlike our algorithm, can handle negative edge lengths.) Finally, we have generalized our algorithm for a version of the dynamic fixed point problem.

Chapter 7
Incremental Algorithms for the Circuit Value Annotation Problem

> *If you see several plans, none of them too sure, if there are several roads diverging from the point where you are, explore a bit of each road before you venture too far along any one—any one could lead you to a dead end.*
> —G. Polya, *Mathematical Discovery, Volume 2*

7.1. Introduction

This chapter presents results on the dynamic circuit value annotation problem. We introduce a new strategy for updating a circuit's annotation incrementally, and show that this strategy yields an exponentially bounded algorithm for the problem. This result, in conjunction with a previous lower bound[Alp90], establishes that the dynamic circuit value annotation problem belongs to the exponentially bounded class. We analyze several variants of this incremental algorithm, show that a version of this algorithm is quadratically bounded for a special class of circuits, and develop bounded versions of this algorithm for the *weighted* circuit value annotation problem. We finally present experimental results showing the practicality of our incremental algorithm.

A *circuit* is a dag in which every vertex u is associated with a function F_u. The output value to be computed at a vertex u is obtained by applying function F_u to the values computed at the predecessors of vertex u. The circuit value annotation problem is to compute the output value associated with each vertex. Thus, the circuit value annotation problem is the problem of computing the unique fixed point of a non-recursive collection of equations. The dynamic version of the problem is to maintain consistent values at each vertex as the circuit undergoes changes [Par83, Rep83, Hoo87, Alp89, Alp90].

From a systems-building perspective, the dynamic circuit value annotation problem is important because it is at the heart of several important kinds of interactive systems, including the pervasive spreadsheet [Bri79, Par83] as well as language-sensitive editors created from attribute-grammar specifications [Rep88]. The dynamic circuit value annotation problem is of interest to incremental computation because the computation performed by an arbitrary program can be represented by a circuit and used in incremental execution of the same program. As Hoover says:

> If, as a consequence of a modification, the results of all intermediate computations change, then the problem is not incremental in nature. $\cdots$ If a large number of the computed results stay the same, however, these computed values can be saved and reused to compute the result after the modification. We do this by constructing a directed graph whose vertices represent the values of computations that are likely to remain unchanged, and whose edges represent the dependencies among these com-

putations. We store the most recently computed value of the computation at the vertex. This graph is called a *dependency graph*, $\cdots$ [Hoo87]

In the case of interactive systems based on attribute grammars, specialized algorithms have been devised that take advantage of the special structure of the problem [Rep83, Rep84, Yeh83, Rep88]. However, a generalized framework has been proposed by Alpern *et al.* that uses the annotation of graphs as a paradigm for specifying other classes of interactive systems, especially ones that cannot be encoded efficiently with attribute grammars [Alp89]. Systems created using this paradigm can give rise to arbitrary circuits. Similarly, Alphonse [Hoo92], a system for automatically generating efficient incremental systems from simple exhaustive imperative program specifications, makes use of incremental algorithms for the general circuit value annotation problem. Thus, the dynamic circuit value annotation problem is highly relevant to real-world systems.

Alpern *et al.* [Alp90] show that the incremental circuit value annotation problem has a lower bound of $\Omega(2^{||\delta||})$ under a certain model of incremental computation called *local persistence*. (This model of computation is discussed in chapter 8.) This chapter outlines bounded algorithms for various versions of the dynamic circuit value annotation problem. In particular, we present an algorithm for the incremental circuit value annotation problem that runs in time $O(2^{||\delta||})$, under the assumption that the evaluation of each function F_u takes unit time. Dropping this assumption leads to the *weighted circuit value annotation problem*, which is also discussed in this chapter.

What is the significance of the results presented in this chapter? First, these results are interesting from the point of view of analyzing the complexity of incremental algorithms in terms of the parameter $||\delta||$. The results show that the circuit value annotation problem has a bounded incremental algorithm—previous to our work, no bounded algorithm for the incremental circuit value annotation problem was known, though several unbounded incremental algorithms were known [Alp90]. Further, the matching exponential lower and upper bounds for this problem establish that the problem is an inherently exponentially bounded problem and enriches the complexity hierarchy for incremental computation.

Second, the algorithms outlined in this chapter appear to be useful in practice. At first blush, these algorithms may appear impractical because they are *exponentially* bounded, with a worst-case complexity of $O(2^{||\delta||})$. In particular, since $||\delta||$ can be $\Omega(n)$ in the worst case (assuming bounded degree graphs for simplicity), it might appear that the incremental algorithm can take time $\Omega(2^n)$ in the worst case. Since a batch algorithm can evaluate the whole circuit in time $O(n)$, again assuming bounded degree graphs, the incremental algorithm may appear inferior to the batch algorithm. However, one version of the algorithm we present runs in time linear in the number of vertices it visits—thus, its worst-case running time is $\Theta(n)$. Hence, from the conventional perspective, the incremental algorithm is asymptotically no worse than the batch algorithm or the nullification-reevaluation incremental algorithm (see Section 7.2) for this problem, while, from the boundedness perspective, the algorithm is better than the batch algorithm and the nullification-reevaluation incremental algorithm.

We present experimental results that show that a version of this incremental algorithm performs well for the attribute updating problem in a Pascal editor, often performing close to the minimum number of evaluations necessary—that is, it runs in time $O(\|\delta\|)$ most of the time.

Third, it can be shown that for a special class of circuits, *monotonic circuits*, and unit changes, the algorithm we present runs in time $O(\|\delta\|^2)$.

The rest of the chapter is organized as follows. Section 2 discusses various basic approaches to the incremental circuit value annotation problem, such as the nullification-reevaluation strategy and various change propagation strategies, and the problems associated with these approaches. Section 3 introduces a new strategy, the iterative evaluate-and-expand strategy, for this problem. Section 4 presents specific bounded algorithms that follow this strategy. Section 5 concerns a generalization of the algorithms discussed in Section 4 to handle the weighted circuit value annotation problem. Section 6 presents experimental results on the performance of a version of these incremental algorithms.

7.2. The Change Propagation Strategy

Let us begin with some relevant terminology.

A circuit is a dag in which every vertex is associated with a function together with an an ordering relation on the predecessors of a vertex. Consider a circuit whose vertices are annotated with (output) values. Let $u.value$ denote the value annotating vertex u. Vertex u is said to be *consistent* if its value equals function F_u applied to the values associated with its predecessor vertices. In other words, if the predecessors of vertex u are $v_1, \ldots, v_k$ in that order, then u is said to be consistent if:

$$u.value = F_u(v_1.value, \ldots, v_k.value).$$

The circuit is said to be *correctly annotated* if each vertex in the circuit is consistent. A vertex is said to be *correct* if its value is the one it would have in a correct annotation of the circuit. Note that a consistent vertex might be incorrect (but only if at least one of its predecessors is incorrect). A change to the circuit consists of the insertion and/or deletion of vertices and/or edges from the dag, along with the associated modification of the functions associated with the various vertices. The changes can leave several of the vertices in the circuit inconsistent, and an incremental algorithm for the circuit value annotation problem needs to update the annotation of the circuit so that all the vertices are consistent. In the algorithms we present in this chapter we will not bother about the exact change to the circuit. Instead, we will assume that an annotated circuit G is given along with a list of the vertices in the circuit that are (possibly) inconsistent—in this chapter we will refer to these vertices as the "modified" vertices—from which a correctly annotated circuit is to be computed. A unit change to the circuit is one that modifies at most one vertex.

Let us now consider the batch version of the circuit value annotation problem. Given an unannotated circuit, we can compute a correct annotation of the circuit by visiting the vertices in the circuit in a topological order [Knu73] and evaluating them. (By "evaluating" a vertex u we mean applying the function associated with vertex u

to the output values currently associated with the predecessors of u, and making the value so obtained the output value of vertex u.) This process has been called topological evaluation [Rep84]. This takes time linear in the number of vertices and edges in the circuit, assuming that the evaluation of each function takes constant time.

The incremental circuit value annotation problem has been studied widely before [Par83, Rep83, Hoo87, Alp90]. It is worth examining the basic approaches that have been used for this problem in previous work.

One simple strategy for computing a correct annotation of an inconsistent circuit is a two-phase algorithm known as the nullification-reevaluation algorithm [Rep84]. In the first phase of this algorithm, the region "downstream" of the modified vertices—that is, the set of all vertices reachable from the modified vertices—is determined using a simple graph traversal, and the values associated with the vertices in this region are "nullified". In the second phase the values of the nullified vertices are evaluated in a topological order. Thus, this algorithm runs in time linear in the size of region downstream of the modified vertices, and is an improvement upon the batch algorithm.

The drawback with the nullification-reevaluation algorithm is that it can perform many unnecessary computations—it is an unbounded algorithm. Note that it is necessary to evaluate a vertex only if it is either a modified vertex or the successor of an affected vertex—otherwise, the vertex must be consistent and re-evaluation will not change its value. "Change propagation" denotes the propagation of the effects of the input change: the propagation starts from the modified vertices, and proceeds downstream until all changes quiesce. It is the basic strategy used in various incremental algorithms for the circuit value annotation problem that attempt to avoid examining that whole of the region downstream of the modified vertices. These algorithms repeatedly choose some vertex for re-evaluation and compute a new value for that vertex from the values of its predecessors until all vertices are consistent. The algorithms differ usually in how they choose vertices to be evaluated. The naive change propagation algorithm presented in Figure 7.1 uses no special strategy in choosing vertices for re-evaluation.

If a change propagation algorithm is to avoid unnecessary vertex evaluations it must ensure that vertices are visited and evaluated in a topological order. For example, consider the circuit shown in Figure 7.2. Labels attached to the edges denote the output value associated with the source vertex of that edge. Expressions inside the vertex denote the function associated with that vertex. The values to the right of the vertex denote the output values associated with that vertex. Vertex u in the circuit is associated with a constant value, which is changed from 1 to 2. This change in input affects the output value of only two vertices, namely u and v. However, after changing the output value associated with vertex u, the naive change propagation algorithm has to choose a vertex from among v and w to re-evaluate. If the algorithm wrongly chooses vertex w to re-evaluate it will temporarily assign a wrong value to that vertex, because it uses v's old (wrong) value of 0 and u's new (correct) value of 2 in computing w's new value. This can cause a "spurious change" to propagate down the circuit. In particular, if the algorithm always chooses the vertex in

procedure NaiveChangePropagation (G, U)
declare
        G : an annotated circuit
        U : a set of vertices in G
        WorkSet : a set of vertices
        u : a vertex
preconditions
        Every vertex in V(G)−U is consistent
begin
[1]   WorkSet := U
[2]   while WorkSet ≠ ∅ do
[3]   Choose and remove some vertex u from WorkSet
[4]   previousValue := u.value
[5]   Re-evaluate u.value from the values of u's predecessors
[6]   if u.value ≠ previousValue then Add the successors of u to WorkSet fi
[7]   od
end
postconditions
        Every vertex in G is consistent

Figure 7.1. A naive change propagation algorithm for the incremental circuit value annotation problem.

workset that is farthest down in the figure it will take time exponential in the number of vertices in the circuit before the spurious change quiesces and the circuit is correctly annotated. (See [Rep84].) On the other hand, an algorithm that correctly chooses to re-evaluate v before w will not erroneously change w's values, and will, consequently, re-evaluate only the vertices u, v, and w and finish the updating in O(1) time.

 If the updating algorithm is to visit and evaluate vertices in a topological order then it is necessary to dynamically maintain a topological ordering of the dag, or equivalent auxiliary information that provides information about the transitive dependences among vertices. For the circuits that arise in the attribute evaluation problem in language-sensitive editors, specialized algorithms have been devised that maintain such auxiliary information without any asymptotic increase in the time complexity of the incremental algorithm [Rep83, Rep84, Yeh83, Rep88]. These algorithms run in time $O(\|\delta\|)$ and are, hence, asymptotically optimal.

 For the general circuit value annotation problem, the best known algorithm that follows this approach is that of Alpern et al.[Alp90]. A dag is said to be correctly prioritized if every vertex u in the dag is assigned a priority, denoted by priority(u), such that if there is a path in the dag from vertex u to vertex v then priority(u) < priority(v). Alpern et al. outline an algorithm for the problem of maintaining a correct prioritization of a circuit as it undergoes modifications. They utilize the priorities in propagating changes in the circuit in a topological order. Thus, in

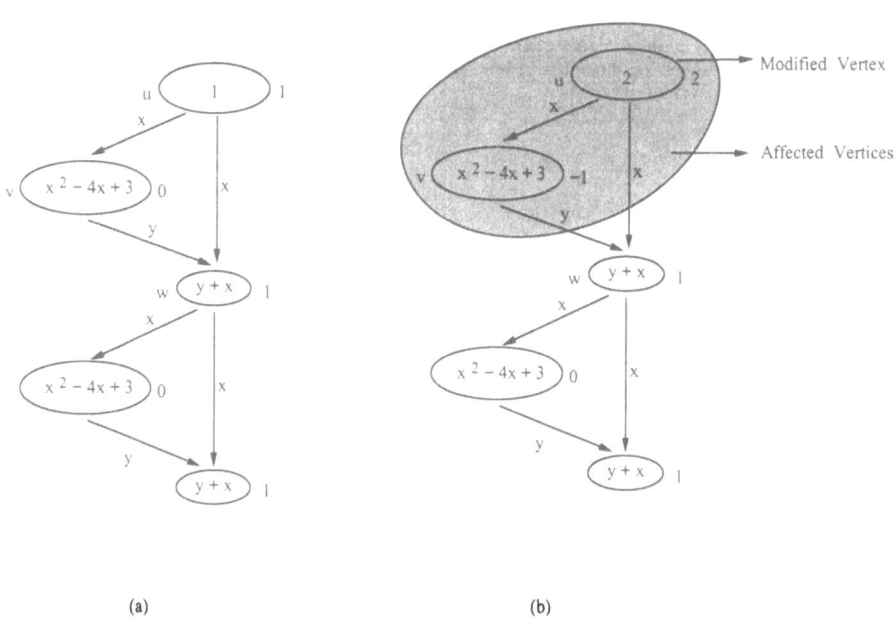

(a)   (b)

Figure 7.2. An example to show the problem with naive change propagation. The circuit shown in (a) is a correctly annotated circuit. The constant-valued function associated with vertex u is modified. The correct annotation of the resulting circuit is shown in (b). Only u and v are affected by this change in the input. But the naive change propagation algorithm can potentially visit all vertices that are reachable from u and temporarily assign wrong values to them. It can also take time exponential in the number of different vertices it visits before it computes the correct annotation of the circuit.

each iteration of the change propagation algorithm outlined in Figure 7.1, the vertex chosen in line [3] is the vertex in WorkSet with the minimum priority. This, however, leads to an unbounded algorithm for the dynamic circuit value annotation problem, because maintaining a topological ordering or priority ordering of the dag can require time unbounded in terms of $\|\delta\|$—the priority ordering of the vertices can greatly change following an input modification, though none of the output values change. For example, it might be necessary to examine all the vertices that are downstream of the modified vertices updating *priorities* even though the values of those vertices are unaffected. Thus, we cannot afford to maintain priorities or a topological ordering of the vertices of the circuit if we desire a bounded algorithm for the dynamic circuit value annotation problem.

7.3. The Iterative Evaluate-and-Expand Strategy

If we do not maintain a priority-ordering or a topological ordering of the vertices of the circuit, how can we avoid the problems with the naive change propagation? The two problems with naive change propagation that we would like to avoid are: (1) The number of vertices that naive change propagation visits and evaluates is not bounded by any function of the number of affected and modified vertices (2) The number of evaluations that naive change propagation performs can be exponential in the number of vertices it visits.

The algorithms we present in this chapter may be described as *iterative evaluate-and-expand* strategies. The essential characteristic of these algorithms is presented in Procedure *IterativeEvaluationAndExpansion* (see Figure 7.3), which is a general schema for computing a correct annotation of a circuit G, given a possibly incorrect annotation of G, and a set of vertices U that are possibly inconsistent. Thus, *IterativeEvaluationAndExpansion* can handle multiple modifications simultaneously. The algorithm works as follows. It initializes the set WorkSet to consist of the initially inconsistent vertices (lines [2]-[5]). Then, the algorithm iteratively does the following until it can be established that the circuit has been correctly annotated: (1) *Evaluate all* the vertices in WorkSet in a relative topological order (see the following paragraph) (line [8]) and (2) *Expand* the WorkSet by adding one or more vertices to it (lines [14]-[18]). After every evaluation step, the algorithm determines if the circuit is correctly annotated as follows: The set of all vertices in WorkSet that have a value different from their original value is identified (line [10]). These vertices are said to be *apparently affected*—some of these vertices may not be affected but just have a wrong value temporarily assigned to them. The set of all successors of the apparently affected vertices, excluding those vertices already in the workset, is identified (line [11])—these are the newly identified *potentially affected* vertices. The algorithm halts if no new vertex is identified as being potentially affected.

The evaluation step is based on the notion of a relative topological ordering. Let H denote the subgraph of G induced by a set of vertices S. Any topological ordering of H is said to yield a *relative topological ordering* for S. Note in particular that if a vertex u topologically precedes vertex v in G and all paths in G from u to v pass through some vertex not in S, then u need not come before v in a relative topological ordering for S. This is important because, in general, it is not possible to determine an actual topological ordering of a set S (*i.e.*, an ordering that accounts for all paths in G) in time bounded by a function of $|S|$ (or even $\|S\|_i$ for any fixed value of i). In contrast, it is possible to determine a relative topological ordering of the vertices of S in time $O(\|S\|)$.

The algorithms presented in this chapter implement specific strategies for the expansion step. We now establish the correctness and some other properties of *IterativeEvaluationAndExpansion*, which are independent of the specific strategy used for the expansion step.

Let Var_i denote the value of a variable Var at the end of the i-th iteration.

procedure IterativeEvaluationAndExpansion (G, U)
declare
        G : an annotated circuit
        U : a set of vertices in G
        WorkSet, *ApparentlyAffected*, *ToBeIncluded* : sets of vertices
        u, v, w: vertices
preconditions
        Every vertex in $V(G)-U$ is consistent
begin
[1]   /* Initialize WorkSet */
[2]   WorkSet := U
[3]   **for** every vertex $u \in U$ **do**
[4] $u.originalValue := u.value$
[5]   **od**
[6]   **loop**
[7]   /* Evaluate WorkSet */
[8]   **for** every vertex $v \in$ WorkSet in *relative topological order* **do** recompute $v.value$ **od**

[9]   /* Test for termination */
[10]   *ApparentlyAffected* := { $v \in$ WorkSet : $v.value \neq v.originalValue$ }
[11]   *PotentiallyAffected* := $Succ$ (*ApparentlyAffected*) – WorkSet
[12]   **if** *PotentiallyAffected* = $\varnothing$ **then** exit loop **fi**

[13]   /* Expand WorkSet */
[14]   *ToBeIncluded* := any non-empty subset of $V(G)$ – WorkSet
[15]   **for** every vertex $v \in$ *ToBeIncluded* **do**
[16]   Insert v into WorkSet
[17] $v.originalValue := v.value$
[18]   **od**
[19]   **end loop**
end
postconditions
        Every vertex in G is consistent

Figure 7.3. The iterative evaluate-and-expand strategy for the incremental circuit value annotation problem. An expansion strategy specifies how the set *ToBeIncluded* is to be chosen in line [14] and leads to a corresponding incremental algorithm for the circuit value annotation problem.

Proposition 7.1.

        1.) Let G be an annotated circuit and U a set of vertices in G such that every vertex in $V(G)-U$ is consistent. Then, the annotation computed by procedure IterativeEvaluationAndExpansion is correct.

        2.) If WorkSet$_i \supseteq N$(AFFECTED), then the algorithm will exit the loop during the $i+1$-th iteration.

3.) If $WorkSet_i \not\supseteq$ AFFECTED, then at least one of the vertices in $PotentiallyAffected_{i+1}$ is an affected vertex.

Proof.

1.) Consider the circuit as annotated when the procedure terminates. We show that every vertex in the circuit is correctly annotated by induction on the vertices v of G in "topological order": we show for every vertex v in G that, assuming the inductive hypothesis that every predecessor of v in G is correct, v is itself correct.

Let $\overline{WorkSet}$ denote the final value of WorkSet. First consider the case that v is in $\overline{WorkSet}$. Whenever the value of a vertex is recomputed, the vertex becomes consistent. It can subsequently become inconsistent only if the value of some of its predecessor changes because of recomputation. Now, the values for vertices in $\overline{WorkSet}$ have been computed in a relative topological order in line [8]—thus, the value of vertex v was recomputed, and the value of none of its predecessors would have been recomputed after that. It follows that every vertex in $\overline{WorkSet}$ is consistent. It follows that every vertex v in WorkSet is also correct (since its predecessors are correct, according to the inductive hypothesis).

Now consider the case that v is not in $\overline{WorkSet}$. Note that the following condition holds true when the procedure terminates: if w and v vertices such that $w \in \overline{WorkSet}$, $v \notin \overline{WorkSet}$, $w \rightarrow v \in E(G)$, then $w.value = w.originalValue$. (Otherwise, v would have been in $PotentiallyAffected$ in the last iteration, and the algorithm would not have halted after that iteration.) Hence, any predecessor w of v that is in $\overline{WorkSet}$ has the same value as it did originally. Since only the values of vertices in $\overline{WorkSet}$ could have changed, any predecessor of v that is *not* in $\overline{WorkSet}$ has the same value as it did initially. Hence, v and *all* of its predecessors have the same values as they did before the update. Since v was initially consistent (from the precondition of the procedure), it must still be consistent and, hence, correct. It follows that *IterativeEvaluationAndExpansion* computes a correct annotation of the circuit.

2.) Assume that $WorkSet_i \supseteq N(\text{AFFECTED})$. Then, after the recomputation in the $i+1$-th iteration, the circuit must be correctly annotated. This follows by induction: a vertex in $\overline{WorkSet}$ must be correct, just as in (1); any vertex not in $\overline{WorkSet}$ is an unaffected vertex, by assumption, and must be correct since it has its original value.

Hence, $ApparentlyAffected_{i+1}$ must be the same as AFFECTED, and $PotentiallyAffected_{i+1}$ must be empty. Hence, the algorithm halts after the $i+1$-th iteration.

3.) Enumerate the vertices of G in some topological order. Let w be the first affected vertex in this ordering that is not in $WorkSet_i$. Since w is not in $WorkSet_i$ it cannot be a modified vertex (that is, one of the vertices in U). Consequently, w must have a predecessor v that is affected. It follows by induction, as before, that every vertex that precedes w in the topological ordering is correctly annotated. In particular, v must be in $WorkSet_i$ and must have a value different from its original value. Hence, v must be in $ApparentlyAffected_{i+1}$ and w must be in $PotentiallyAffected_{i+1}$. $\square$

Let us now briefly consider the two concerns we raised at the beginning of the section, concerning the number of vertices visited and the number of vertex evaluations performed. The above algorithm performs | WorkSet | evaluations in each iteration, and iterates at most | $\overline{\text{WorkSet}}$ | times, since it adds at least one vertex to WorkSet in each iteration. Consequently, it performs at most | $\overline{\text{WorkSet}}$ |^2 evaluations—that is, the number of evaluations the algorithm performs is, in the worst-case, no more than quadratic in the number of vertices it visits. The above claim is true irrespective of the strategy used to add vertices to the WorkSet. We will show in the next section that a particular strategy of adding vertices to the WorkSet yields an algorithm in which the number of evaluations performed is linear in the number of vertices visited. We will also present strategies to keep the number of vertices visited bounded (by some function of $\| \delta \|$).

Remark 7.2. Note that some improvements are possible to the above algorithm. For instance, instead of evaluating every vertex in WorkSet in line [8] we can do the following: we can determine the set of all vertices in WorkSet that are reachable from the set of vertices that were most recently added to the WorkSet (in line [2] or in lines [15]-[18]) *in the subgraph induced by WorkSet* and evaluate only these vertices—the values of other vertices are guaranteed not to change. We have omitted such improvements to keep the algorithm description simple. These improvements do not, however, change the worst-case time complexity—in the worst-case, the number of vertices evaluated in each iteration will still be $O(|\text{WorkSet}|)$. However, we present experimental results in Section 7.6 which show that the above improvements make a big difference in a practical sense.

We will now see how the number of vertices that the algorithm visits can be bounded by some function of $\| \delta \|$.

7.4. Breadth-First Expansion

7.4.1. Bounded-Outdegree Circuits

We now consider two specific strategies for expansion. Let *BF_Expansion* be the algorithm obtained by refining line 10 of *IterativeEvaluationAndExpansion* to "*ToBeIncluded* := *Succ* (WorkSet) – WorkSet". This algorithm expands WorkSet in a "breadth-first" fashion. At the beginning of the *i*-th iteration, WorkSet consists exactly of all the vertices reachable from some modified vertex along a path consisting of $(i-1)$ or less edges. Let *RBF_Expansion* be the algorithm obtained by refining line 10 of *IterativeEvaluationAndExpansion* to "*ToBeIncluded* := *PotentiallyAffected*". This algorithm performs a "restricted breadth-first" expansion of WorkSet—expansion occurs beyond a vertex only if that vertex appears to be an affected vertex.

We now show that both of the above strategies lead to a bounded incremental algorithm in the case of circuits with bounded outdegree. A *k*-ary circuit is one in which the outdegree of every vertex is less than or equal to *k*. In particular, a binary circuit is a circuit in which the outdegree of every vertex is less than or equal to 2.

Proposition 7.3. Both *BF_Expansion* and *RBF_Expansion* process unit changes to binary circuits in time $O(2^{|\delta|})$. They process unit changes to k-ary circuits in time $O(k^{|\delta|})$. They process arbitrary changes to binary circuits in time $O(|\delta| \cdot 2^{|\delta|})$ and arbitrary changes to k-ary circuits in time $O(|\delta| \cdot k^{|\delta|})$

Proof. We prove the result for binary circuits. The result for k-ary circuits follows similarly. Proposition 7.1.3 implies that both *BF_Expansion* and *RBF_Expansion* add at least one affected vertex to WorkSet in each iteration until all the affected vertices are in WorkSet. It follows that these algorithms make at most $|\text{AFFECTED}|+1$ iterations. Because every vertex in the circuit has outdegree at most 2, at most $|U| \times 2^i$ new vertices can be added to WorkSet during the i-th iteration. Hence, at the beginning of the i-th iteration, $|\text{WorkSet}| \leq \sum_{j=0}^{i-1} |U| \times 2^j = |U| \times (2^i - 1)$. The i-th iteration itself takes time $O(|U| \times 2^i)$.[1] The whole algorithm takes time $O(\sum_{i=1}^{|\text{AFFECTED}+1|} |U| \times 2^i) = O(|U| \times 2^{|\text{AFFECTED}|})$. The result follows. □

In the following sections we examine some refinements and special cases of the algorithms described above.

It follows from the above proof that the above boundedness result holds true for any expansion strategy that (1) adds at least one affected vertex to WorkSet in each iteration until all affected vertices are in WorkSet and (2) multiplies the size of WorkSet by at most a factor of two (or some constant k) in each iteration. If instead of (1) we can guarantee that the expansion strategy adds at least one vertex in $N(\text{AFFECTED})$ to WorkSet in each iteration, then we would still get a bounded algorithm, albeit an algorithm that processes unit changes in time $O(2^{||\delta||})$ instead of time $O(2^{|\delta|})$. This is roughly the idea behind our bounded incremental algorithm for circuits with unbounded outdegree—we can expand in each iteration by adding, say, one successor of every vertex in *ApparentlyAffected* to WorkSet. This idea is explored in the following section.

Note that both *BF_Expansion* and *RBF_Expansion* have the same worst-case complexity and are both bounded incremental algorithms. Since there seems no good reason to add all of *Succ* (WorkSet) to *Workset* instead of just *Succ* (*ApparentlyAffected*), *RBF_Expansion* seems preferable to *BF_Expansion*. But *RBF_Expansion* does not always do better than *BF_Expansion*—while there are classes of input instances where *BF_Expansion* can take time that is an exponential function of the time *RBF_Expansion* takes, there are also classes of input instances where *RBF_Expansion* can take time that is a quadratic function of the time that *BF_Expansion* takes. Consider, for example, the circuits shown in Figure 7.4. In both cases consider a modification that changes the constant value associated with the "root" of the circuit from 1 to 2. For the circuit on the left, *BF_Expansion* visits all

---

[1]Note that a relative topological ordering of a set S of vertices can be determined in time $O(|S|)$ in the case of binary circuits.

the vertices, while *RBF_Expansion* visits only the vertices in the shaded region. In particular, if h is the height of the circuit, then *BF_Expansion* runs in time $O(h^2)$, while *RBF_Expansion* runs in time $O(2^h)$. In contrast, both *BF_Expansion* and *RBF_Expansion* visit all vertices in the circuit on the right. Let h denote the height of the circuit and n the number of vertices in the circuit. However, while *BF_Expansion* finishes the updating in approximately h iterations, *RBF_Expansion* takes approximately 2^h iterations. Thus, *BF_Expansion* takes $O(n)$ time for the updating, while *RBF_Expansion* takes $O(n^2)$ time for the updating.

Both *BF_Expansion* and *RBF_Expansion* have their advantages. We show in Section 7.4.3 that for a particular class of circuits, *RBF_Expansion* is polynomially bounded. We present in Section 7.4.4 an adaptation of *BF_Expansion* that, unlike *RBF_Expansion*, runs in time linear in the number of vertices it visits.

7.4.2. Handling Unbounded Outdegree

Note that the breadth-first expansion strategy does not yield a bounded incremental algorithm for circuits in which the outdegrees of vertices cannot be bounded by some constant. The reason is that in procedure *RBF_Expansion*, an unaffected vertex z,

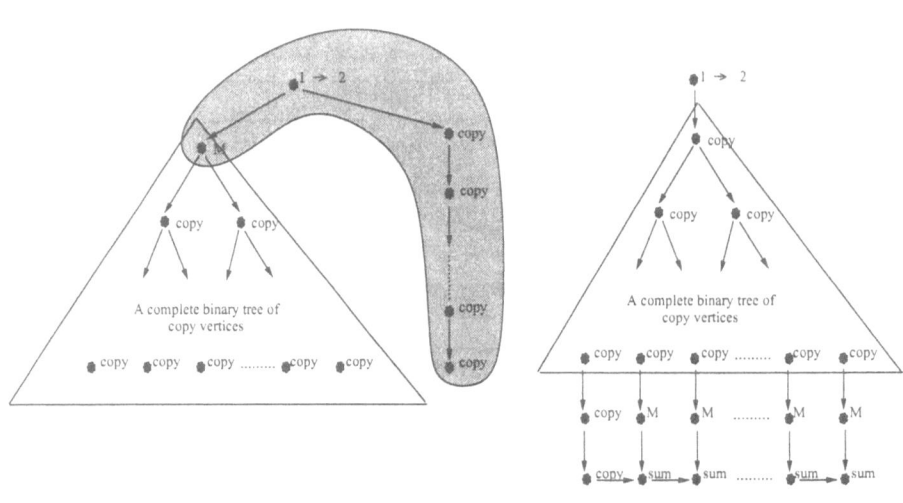

Figure 7.4. Example circuits for comparing the performance of *BF_Expansion* and *RBF_Expansion*. Vertices labelled *copy* just copy their input to their output—consequently they propagate any change in their input to their output. Vertices labelled M are associated with the function $\lambda x.\ \max(x, 5)$—the examples have been constructed so that the change in the input values of M vertices do not change their output value; consequently, these vertices stop change propagation in the case of *RBF_Expansion*. Vertices labelled *sum* output the sum of their two inputs. The "root" vertices of both circuits are associated with the constant function 1, which is changed to 2.

which by definition is initially correct, may be given an incorrect value at some intermediate iteration i. Although, z's correct value will ultimately be restored by the time
RBF_Expansion terminates, z's successors are part of the WorkSet at the end of itera-
tion i. Because z is not affected $\|\delta\|$ does not include or account for the number of
successors of z. Consequently, the number of successors z has, and the amount of
work the algorithm does, can be arbitrarily large even if $\|\delta\|$ is bounded by a small
constant.

However, the breadth-first expansion strategy can be easily adapted to handle
arbitrary circuits in a bounded fashion. Let G be any circuit. $G*$ is a binary circuit
equivalent to G obtained as follows. (See Figure 7.5.) Let u be a vertex in G with k
successors $v_1, \cdots, v_k$ where $k > 2$. Replace u by $k-1$ vertices $u_1, \cdots, u_{k-1}$ each of
out-degree 2. Vertex u_1 has the same function and the same set of predecessors as
vertex u, and two successors v_1 and u_2. For $1 < i \le k-1$, vertex u_i has a single predecessor u_{i-1}, and is associated with the identity function. The two successors of vertex
u_i, where $1 < i < k-1$, are v_i and u_{i+1}. The two successors of vertex u_{k-1} are v_{k-1} and
v_k. $G*$ is obtained from G by thus duplicating all vertices with outdegree greater than
2.

It is trivial to update the structure of $G*$ as and when vertices/edges are
inserted and deleted from G. For any change δ, $|\delta|_{G*} \le \|\delta\|_G$, and $\|\delta\|_{G*} \le$
$2 \cdot \|\delta\|_G$. Consequently, it is possible to handle circuits with unbounded outdegree
using the vertex duplication scheme outlined above. In reality it is not necessary to
construct and work with the circuit $G*$ described above—we can effectively simulate
the action of the breadth-first expansion strategy on G^*, given just G. We can expand
in each iteration by adding one successor of every vertex in either WorkSet (breadth-
first expansion) or *ApparentlyAffected* (restricted breadth-first expansion) to WorkSet.

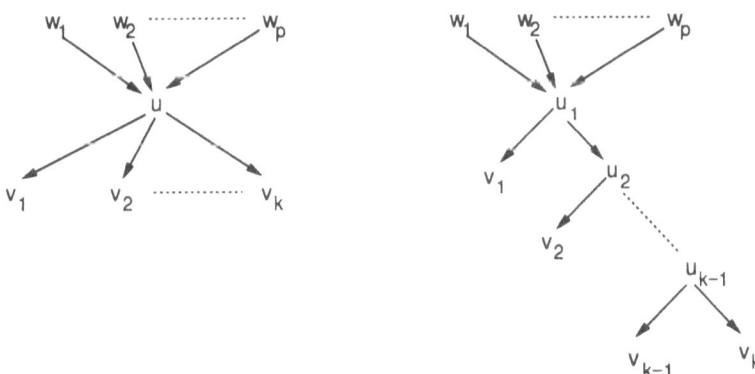

Figure 7.5. Duplicating vertices to transform unbounded outdegree graphs to bounded outdegree graphs.

Proposition 7.4. By using vertex duplication, we can process unit changes in arbitrary circuits in time $O(2^{\|\delta\|})$, and arbitrary changes in arbitrary circuits in time $O(\|\delta\| \cdot 2^{\|\delta\|})$.

7.4.3. Monotonic Circuits: A Special Case

We now show that the running time of *RBF_Expansion* can be bounded by $\|\delta\|^2$ under some circumstances.

Proposition 7.5. Assume that there exists a total order on the set of all (possible) output values such that every function F_u is monotonic with respect to this total order. Then, *RBF_Expansion* processes unit changes (*i.e.*, $|U| = 1$) in time $O(\|\delta\|^2)$.

Proof. Assume, without loss of generality, that the new value of the modified vertex u is greater than its original value. Then, the new value of every affected vertex v must be greater than its (v's) original value. In fact, it follows by induction on the steps performed by the algorithm that the recomputation of the value of any vertex v either leaves its value unaltered or increases it, but never decreases it. Hence, if the value of any vertex v changes (increases) during some particular recomputation step, then no subsequent recomputation will restore v's original value. That is, v must be an affected vertex. Hence, every vertex in WorkSet is either an affected vertex or the successor of an affected vertex. It follows that the algorithm makes at most $\|\delta\|$ iterations. Each iteration takes $O(\|\delta\|)$ time, under the assumption of bounded outdegree. Unbounded outdegree graphs can be handled using vertex duplication. $\square$

Note that the above result does not carry over to the case of non-unit changes (*i.e.*, $|U| > 1$) because, in this case, the values of some of the modified vertices might increase, while the values of the remaining modified vertices decrease. Spurious propagation is possible in this case, resulting in an exponential time complexity. However, if the values of all the modified vertex increase together (or decrease together) then the algorithm does terminate in $O(\|\delta\|^2)$ time.

Note that the monotonic circuit problem is related to some of the problems we have looked at in previous chapters. Consider the SWSF fixed point problem restricted to dags—that is, SWSF fixed point problems for which the dependence graph is acyclic. This problem is a special case of the monotonic circuit problem, since SWSF functions are a special case of monotonic functions. Similarly, consider the longest path problem in dags with one or more source vertices. Again this problem is a monotonic circuit problem, even in the presence of negative length edges, though it is not a SWSF fixed point problem. We have seen in earlier chapters that these problems have an $O(\|\delta\| \log \|\delta\|)$ algorithm. The algorithm we have presented for the general circuit value annotation problem, which assumes nothing about the kind of functions associated with the vertices, turns out to handle these problems reasonably efficiently, in $O(\|\delta\|^2)$ time, at least in the case of unit changes.

7.4.4. A Tradeoff

We observed at the end of Section 7.3 that the number of evaluations performed by any iterative evaluate-and-expand strategy is no worse than a quadratic function of the number of vertices it visits. There are classes of input instances for which this worst-case quadratic behavior is achieved by both *BF_Expansion* and *RBF_Expansion*. For example, consider a chain of vertices (see Figure 7.6(a)) in which the first vertex is modified and all the vertices are affected. Let *n* denote the number of vertices in the chain. Both *BF_Expansion* and *RBF_Expansion* perform *n* iterations. In the beginning of the *j*-th iteration WorkSet consists of the first *j* vertices in the chain, all of which are evaluated. Consequently, the algorithms end up performing $\Theta(n^2)$ work.

This example shows that *BF_Expansion* and *RBF_Expansion* can end up doing more work than the nullification-reevaluation algorithm, which, since it does work linear in the number of vertices it visits, can process the example in Figure 7.6(a) in $O(n)$ time. Of course, the nullification-reevaluation algorithm will often visit many more vertices than either *BF_Expansion* or *RBF_Expansion*. What we see here is a trade-off between the number of vertices visited and the number of vertex evaluations performed: in trying to bound the number of vertices visited by a function

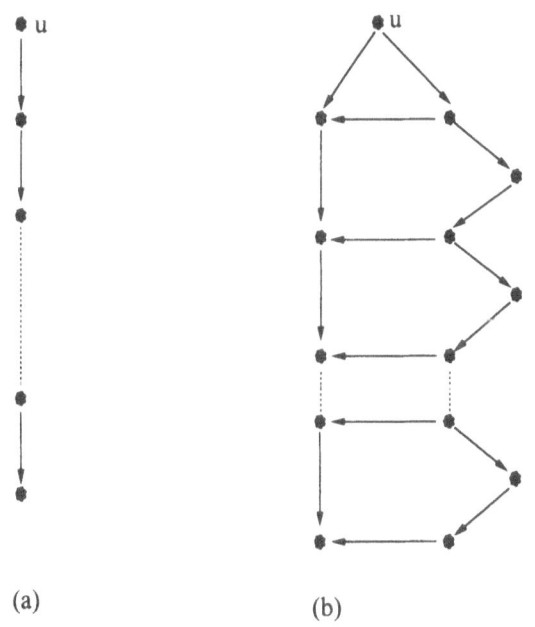

(a)   (b)

Figure 7.6. The above examples show one of the disadvantages of the original breadth-first expansion.

of $\| \delta \|$ we have potentially increased the number of times each vertex is evaluated. We will now see how we can reduce the cost of repeated evaluations.

The expansion algorithms exhibit the above kind of behavior usually in circuits which have long chains. In such cases, the cost of repeated evaluations completely dominates the cost of expansion. A simple improvement that can often avoid this kind of behavior is the improvement outlined in Remark 7.2. With this improvement, the expansion algorithms can process the example in Figure 7.6(a) in $O(n)$ time, since in each iteration only the vertex most recently added to the WorkSet is evaluated. However, this does not fix the problem completely. Consider the second example in Figure 7.6. If the functions associated with the vertices are appropriately defined, the expansion algorithms will end up evaluating $\Theta(j)$ vertices in the j-th iteration, and do $\Theta(n^2)$ work overall.

We can, in general, reduce the overhead of repeated evaluations if we choose to *repeatedly* expand until the size of WorkSet doubles before we initiate a evaluation step. This idea leads to the algorithm *Balanced_BF_Expansion* presented in Figure 7.7, a variant of *BF_Expansion* that runs in time linear in the number of vertices it visits.

We briefly explain the algorithm. Recall that in a breadth-first expansion step we add the vertices in *Succ* (WorkSet) to WorkSet. Let us denote by *Fringe* the set of vertices most recently added to WorkSet, that is, in the most recent expansion step. Obviously, in an expansion step, it is sufficient to add *Succ* (*Fringe*) to WorkSet, since *Succ*(WorkSet−*Fringe*) is already contained in WorkSet.

Let us relate the behavior of *BF_Expansion* to the behavior of *Balanced_BF_Expansion*. Let us denote the set $X \cup Succ(X)$ by $F(X)$. Let $F^0(X)$ denote X, and let $F^{i+1}(X)$ denote $F(F^i(X))$. *BF_Expansion* computes and evaluates all the vertices in $F^0(U)$, $F^1(U)$, $F^2(U)$, $\cdots$, until it reaches the first $F^p(U)$ that contains AFFECTED $\cup$ *Succ* (AFFECTED) at which point the algorithm terminates. *Balanced_BF_Expansion* too computes this sequence of sets, but does not perform a evaluate step for each of these sets. Instead, it evaluates the sequence of sets $F^{i_0}(U)$, $F^{i_1}(U)$, $\cdots$, $F^{i_q}(U)$, where i_j is the minimum k such that[2] $|F^k(U)| \geq |U| \times 2^j$. The algorithm terminates after it performs an evaluation of the first set $F^{i_q}(U)$ that completely contain AFFECTED $\cup$ *Succ* (AFFECTED). The requirement on the cardinality of $F^{i_q}(U)$ implies that i_q may be greater than p (recall that $F^p(U)$ is the last set evaluated by *BF_Expansion*) and hence *Balanced_BF_Expansion* may visit and evaluate more vertices than *BF_Expansion*. But observe that $i_{q-1} < p \leq i_q$. Consequently, we can place a bound on the extra number of vertices that *Balanced_BF_Expansion* visits. We show below that the number of vertices in $F^{i_q}(U)$ is at most four times the number of vertices in $F^{i_{q-1}}(U)$. It follows that

---

[2]This condition may fail to hold at a boundary situation, when all successors of vertices in WorkSet are already in the WorkSet, and WorkSet cannot be expanded any more. But this can only happen in the very last iteration.

Balanced_BF_Expansion visits at most four times the number of vertices *BF_Expansion* visits. We will also show that the time complexity of *Balanced_BF_Expansion* is linear in the number of vertices it visits. Consequently, *Balanced_BF_Expansion* never does more than a constant times the work that *BF_Expansion* does, though *BF_Expansion* can, in a number of cases, do work that is quadratic in the work that *Balanced_BF_Expansion* does.

Proposition 7.6. If the input circuit is a binary circuit, then the following invariant always holds true at line [18] of procedure *Balanced_BF_Expansion*: $|Fringe| \leq |U| \times 2^{j-1} \leq |WorkSet| < |U| \times 2^j$. (Observe that j is the loop count.)

Proof. The claim is true the very first time execution reaches line [18], since both WorkSet and *Fringe* are then equal to U and j is 1. Now we consider two cases.

Let us now assume that the invariant holds true at line [18] at some point. Assume that lines [18]-[24] are then executed, and that the termination condition in line [25] fails to hold, and execution returns to line [18]. The lower bound on the cardinality of WorkSet will still hold trivially, since we have only added vertices to WorkSet. The upper bound on WorkSet's cardinality also holds, since otherwise the repeat loop would have terminated. Note that at the end of executing lines [18]-[24] *Fringe* denotes exactly the set of vertices that were added to WorkSet in these lines. Consequently, the cardinality of *Fringe* must be exactly the increase in the cardinality of WorkSet. Since the cardinality of WorkSet was at least $|U| \times 2^{j-1}$ at the beginning and is less than $|U| \times 2^j$, it follows immediately that the upper bound on the cardinality of *Fringe* also holds.

We now consider the case when line [18] is executed the first time in the j-th iteration of the outer loop (lines [8]-[26]) for some $j > 1$, We assume that the invariant was true the previous time line [18] was executed, after which lines [19]-[24] were executed, and the termination condition in line [25] was found to be true. The lines [18]-[24] replace *Fringe* with *Succ(Fringe)*–WorkSet. Since we are dealing with binary circuits, this can multiply the cardinality of *Fringe* by at most two. Consequently, the lower bound on the cardinality of *Fringe* continues to hold (since the value of j has increased by one meanwhile). The lower bound on the cardinality of WorkSet follows trivially from the termination condition in line [25]. The upper bound on the cardinality of WorkSet follows from the previous upper bound and the upper bound on $|Fringe|$, which is the set of vertices that were added to WorkSet. □

Assume that the algorithm terminates after k iterations of the outer loop. Note that the above algorithm too is a refinement of the original scheme outlined in Figure 7.3. Consequently, it follows from Proposition 7.1c that that the algorithm adds at least one affected vertex to WorkSet in each iteration of the outer loop except possibly the last one. Hence, we have $k \leq |AFFECTED|+1$. The work done in the i-th iteration can be bounded by the cardinality of WorkSet at the end of the i-th iteration. Consequently, the work done in the i-th iteration is $\Theta(|U| \times 2^i)$, and the algorithm terminates in time $O(|U| \times 2^k)$, which is $O(|U| \times 2^{|AFFECTED|})$. Further, the cardinality of WorkSet, when the algorithm halts, is $\Theta(|U| \times 2^k)$, from the above Proposition. Hence, this algorithm runs in time linear in the number of vertices it visits.

---

procedure Balanced_BF_Expansion (G, U)
declare
        G : an annotated circuit
        U : a set of vertices in G
        WorkSet, *ApparentlyAffected, Fringe, ToBeIncluded* : sets of vertices
        u, v, w: vertices
preconditions
        Every vertex in $V(G)-U$ is consistent
begin
[1]   /* Initialize WorkSet and Fringe */
[2]   WorkSet := U
[3]   *Fringe* := U
[4]   **for** every vertex $u \in U$ **do**
[5] $u.originalValue := u.value$
[6]   **od**
[7]   j := 0

[8]   **loop**
[9]   j := j+1
[10]   /* Evaluate WorkSet */
[11]   **for** every $v \in$ WorkSet in *relative topological ordering* **do** recompute $v.value$ **od**

[12]   /* Test for termination */
[13]   *ApparentlyAffected* := { $v \in$ WorkSet : $v.value \neq v.originalValue$ }
[14]   *PotentiallyAffected* := *Succ* (*ApparentlyAffected*) − WorkSet
[15]   **if** *PotentiallyAffected* = $\emptyset$ **then** exit loop **fi**

[16]   /* Iteratively expand WorkSet until its cardinality is at least $2^j \times |U|$
[17]   * unless there are no more vertices to add to WorkSet */
[18]   **repeat**
[19]   *ToBeIncluded* := *Succ* (*Fringe*) − WorkSet
[20]   **for** every vertex $v \in$ *ToBeIncluded* **do**
[21]   Insert v into WorkSet
[22] $v.originalValue := v.value$
[23]   **od**
[24]   *Fringe* := *ToBeIncluded*
[25]   **until** | WorkSet | $\geq |U| \times 2^j$ **or** *Succ* (WorkSet) $\subseteq$ WorkSet
[26]   **end loop**
end
postconditions
        Every vertex in G is consistent

Figure 7.7. Procedure Balanced_BF_Expansion balances the work done in the expansion step with the work done in the evaluation step. Consequently, it runs in time linear in the number of vertices it visits.

Since the vertices it visits are all downstream of modified vertices, this algorithm is strictly better than the nullification-reevaluation algorithm.

7.5. The Weighted Circuit Value Annotation Problem

In general, it is not true that each function F_u in a circuit can be computed in unit time. For instance, it might be necessary to look at the values of all the predecessors of vertex u in order to compute the value at u. In this case, it might be more reasonable to assume that the cost of computation of F_u is proportional to the indegree of vertex u. In this section we consider the weighted version of the incremental circuit value annotation problem, where the cost of computation of function F_u is an arbitrary, but known, value ≥ 1, denoted by $u.cost$. We refer to $u.cost$ as the cost of vertex u. If X is a set of vertices, then $Cost(X)$ is defined to be the sum of the costs of the vertices in X. We will denote $Cost(\text{CHANGED} \cup Succ\,(\text{AFFECTED}))$ by C_δ. We present a bounded-cost scheduling algorithm for the circuit value annotation problem in this section. As before, the outlined algorithm works for binary circuits, and general circuits may be handled using vertex duplication.

The breadth-first expansion does not yield a bounded algorithm for the weighted circuit value annotation problem. It ensures that the *number* of vertices visited is bounded by an exponential function of $\|\delta\|$. But the *cost of recomputation* of the values of the visited vertices that are not in $N(\text{AFFECTED})$ might be unbounded in terms of $\|\delta\|$ or C_δ. In order to bound the cost of recomputation of the values of the visited vertices by some function of $\|\delta\|$ and C_δ, we essentially use a weighted breadth-first expansion strategy (*i.e.*, a shortest-path-first expansion strategy), where the weight of each vertex is the cost of recomputation of that vertex's value.

More formally, the algorithm (see Figure 7.8) works as follows. For any vertex v reachable from some modified vertex, define $v.leastPathCost$ to be the minimum over all paths from some modified vertex to vertex v of the path's cost. (The cost of a path is the sum of the costs of the vertices in the path.) Initially WorkSet consists of the set of all modified vertices (*i.e.*, U). During every iteration, the algorithm chooses the vertex v not in WorkSet for which $v.leastPathCost$ is minimum and adds it to WorkSet.[3] This leads to an $2^{2 \cdot C_\delta}$ algorithm for binary circuits, as explained below.

Let $\overline{\text{WorkSet}}$ denote the final value of WorkSet. We first obtain a bound on $Cost(\overline{\text{WorkSet}})$.

Proposition 7.7. $Cost(\overline{\text{WorkSet}}) \leq |U| \cdot 2^{C_\delta}$.

---

[3]Observe that there may be several vertices outside WorkSet whose *leastPathCost* is minimum. We can choose any one these vertices and add it to WorkSet. Alternatively, we could add *all* the vertices with minimum *leastPathCost* to WorkSet simultaneously. This would make the algorithm look more like *BF_Expansion*, but does not affect any of the following complexity analysis.

procedure WeightedExpansion (G, U)
declare
      G : an annotated circuit
      U : a set of vertices in G
      WorkSet, *ApparentlyAffected*, *PotentiallyAffected*: sets of vertices
      u, v, w: vertices
preconditions
      Every vertex in $V(G)-U$ is consistent
      $v.leastPathCost$ is ∞ for every vertex v
begin
[1]   /* Initialize WorkSet */
[2]   WorkSet := U
[3]   **for** every vertex $u \in U$ **do**
[4] $u.originalValue := u.value$
[5] $u.leastPathCost := u.cost$
[6]   **for** every vertex $w \in Succ(u)$ **do**
[7] $w.leastPathCost := \min(w.cost+u.leastPathCost, w.leastPathCost)$
[8]   **od**
[9]   **od**

[10]   **loop**
[11]   /* Evaluate WorkSet */
[12]   **for** every vertex $v \in$ WorkSet in *relative topological order* **do** recompute $v.value$ **od**

[13]   /* Test for termination */
[14]   *ApparentlyAffected* := { $v \in$ WorkSet : $v.value \neq v.originalValue$ }
[15]   *PotentiallyAffected* := $Succ$ (*ApparentlyAffected*) - WorkSet
[16]   **if** *PotentiallyAffected* = $\varnothing$ **then** exit loop **fi**

[17]   /* Expand WorkSet: identify the vertex in $Succ$ (*WorkSet*)–*WorkSet* whose leastPathCost
[18]   * is minimum and add it to WorkSet. If there are several minimum vertices, any one can
[19]   * be added to *Workset*. Alternatively, all minimum vertices can be added to *Workset*. */
[20]   Choose a vertex v from $Succ$ (*WorkSet*) – WorkSet for which $v.leastPathCost$ is minimum
[21]   Insert v into WorkSet
[22] $v.originalValue := v.value$
[23]   **for** every vertex $w \in Succ(v)$ **do**
[24] $w.leastPathCost := \min(w.cost+v.leastPathCost, w.leastPathCost)$
[25]   **od**
[26]   **end loop**
[27]   **for** every vertex $v \in$ (WorkSet $\cup$ $Succ$ (WorkSet)) **do** $v.leastPathCost := \infty$ **od**
end
postconditions
      Every vertex in G is consistent
      $v.leastPathCost$ is ∞ for every vertex v

Figure 7.8. An incremental algorithm for the weighted circuit value annotation problem.

Proof. It is possible to construct a subgraph F of G, a shortest-path forest, which satisfies the following condition. F is a forest, consisting of one tree with root u for each vertex u in U. The set of vertices in F, $V(F)$, is WorkSet. For any vertex v in a tree with root u, $v.leastPathCost = Cost(P)$, where P is the unique path from u to v in F.

Let w denote the last vertex to be added to WorkSet. Then, $v.leastPathCost \le w.leastPathCost$, for every vertex v in WorkSet. It follows from Proposition 7.1.3 that w must be a vertex in $N(\text{AFFECTED})$. Hence, there exists a path P from some vertex in U to w consisting only of affected vertices (except possibly for w itself). It follows that $w.leastPathCost \le Cost(P) \le C_\delta$. Thus, we observe that for every vertex v in the forest F, $v.leastPathCost \le C_\delta$.

Let T be a binary tree, every vertex of which has an associated integer cost ≥ 1. If v is a vertex in the tree, define $v.leastPathCost$, as above, to be the sum of the costs of vertices on the path from the root of the tree to v (including the endpoints). T is said to be a k-tree if for every vertex v in T, $v.leastPathCost \le k$. If T is a k-tree then, $Cost(T)$, the sum of the costs of the vertices in T, is bounded by $2^k - 1$. This can be established as follows. A tree is a k-tree iff the root has cost $i \le k$, and each of the root's subtrees are $(k-i)$-trees. Let $C(k)$ denote the maximum over all k-trees T of $Cost(T)$. Then,

$$C(0) = 0;$$
$$C(1) = 1;$$
$$C(k) = \max_{1 \le i \le k} (i + 2 \times C(k-i)), \text{ for } k \ge 0.$$
$$= 1 + 2 \times C(k-1)), \text{ for } k \ge 0.$$

It follows immediately that $C(k)$ is $2^k - 1$. (A similar result holds even if vertex costs are real, instead of being integers.)

It follows that $Cost(T) \le 2^{C_\delta}$ for every tree T in the forest F. The number of trees in F is $|U|$. Hence, $Cost(\text{WorkSet})$ is $O(|U| \cdot 2^{C_\delta})$. □

Proposition 7.8. *WeightedExpansion processes unit changes in binary circuits in time* $O(2^{2 \cdot C_\delta})$. *It processes arbitrary changes in binary circuits in time* $O(|\delta|^2 \cdot 2^{2 \cdot C_\delta})$. *Thus, by using vertex duplication, we can process unit changes in arbitrary circuits in time* $O(2^{2 \cdot (C_\delta + \|\delta\|)})$, *and arbitrary changes in arbitrary circuits in time* $O(\|\delta\|^2 \cdot 2^{2 \cdot (C_\delta + \|\delta\|)})$.

Proof. We first establish the result for binary circuits. The time taken to recompute values of vertices in WorkSet is $O(Cost(\text{WorkSet}))$. Hence, a single iteration of the loop in lines [10]-[26] takes times $O(Cost(\text{WorkSet}))$, which is bounded by $O(Cost(\text{WorkSet}))$. The algorithm makes at most $|\text{WorkSet}|$ iterations. Since $|\text{WorkSet}| \le Cost(\text{WorkSet})$ the algorithm runs in time $O(Cost(\text{WorkSet})^2)$. Hence, the result follows for binary circuits.

The result for arbitrary circuits follows from the technique of vertex duplication. Note that when using vertex duplication, the cost of computing the identity function should be taken to be 1. It can be verified that when a unbounded circuit G is converted into a binary circuit G^* using vertex duplication C_δ in G^* is bounded by $C_\delta + \|\delta\|$ in G. The proposition follows. □

More efficient weighted expansion

We observed earlier that *BF_Expansion* could be improved by balancing the work done by the expansion and evaluation steps. Similarly, the weighted expansion algorithm too can be made more efficient as follows. Let $\overline{\text{WorkSet}}$ denote the the final value of WorkSet in the above algorithm. Consider the case of unit changes in binary circuits. The above algorithm can potentially take time $\Theta(2^{2 \cdot C_\delta})$ to process a change because that algorithm recomputes the values of vertices in WorkSet every time a vertex is added to WorkSet. The total time spent on recomputation of values becomes quadratic in $Cost(\overline{\text{WorkSet}})$, which is $\Theta(2^{C_\delta})$ in the worst case. If we avoid recomputing values every time a vertex is added to WorkSet, and instead recompute values every time $Cost(\text{WorkSet})$ (roughly) doubles, then the total cost of recomputation will be linear in $Cost(\overline{\text{WorkSet}})$. This idea leads to the algorithm outlined in Figure 7.9. However, the time complexity of this algorithm is not $O(Cost(\overline{\text{WorkSet}}))$. In the worst case, the dominating factor in the time complexity of the new algorithm is the time spent on choosing the vertex to add to WorkSet in each iteration. This will cost $O(|\overline{\text{WorkSet}}| \log |\overline{\text{WorkSet}}|) = O(C_\delta \cdot 2^{C_\delta})$.

Proposition 7.9. *ImprovedWeightedExpansion* processes unit changes in binary circuits in time $O(C_\delta \cdot 2^{C_\delta})$. It processes arbitrary changes in binary circuits in time $O(C_\delta^2 \cdot 2^{C_\delta})$. Thus, by using vertex duplication, we can process unit changes in arbitrary circuits in time $O((\|\delta\| + C_\delta) \cdot 2^{2 \cdot (\|\delta\| + C_\delta)})$, and arbitrary changes in arbitrary circuits in time $O((\|\delta\| + C_\delta)^2 \cdot 2^{2 \cdot (\|\delta\| + C_\delta)})$.

Proof. Let $\overline{\text{WorkSet}}$ denote the final value of WorkSet as computed by the We show that the amount of time spent by *ImprovedWeightedExpansion* on re-evaluations of values of vertices is $O(Cost(\overline{\text{WorkSet}}))$. The algorithm makes $O(|\overline{\text{WorkSet}}|)$ priority queue operations in choosing vertices to add to WorkSet, each operation taking time $O(\log|\overline{\text{WorkSet}}|)$. Hence, the whole algorithm runs in time $O(Cost(\overline{\text{WorkSet}}) \log Cost(\overline{\text{WorkSet}}))$.

Note that $WorkSetCost_i$, abbreviated W_i, is the sum of the costs of the vertices in WorkSet at the beginning of the $i+1$-th iteration, while $totalCost_i$, abbreviated T_i, is $\sum_{j=0}^{i-1} WorkSetCost_j$. Thus, while W_i is the time spent on re-evaluation in the $i+1$-th iteration, T_i is the time spent on re-evaluations in the first i iterations. Assume that the procedure makes k iterations of the outer loop. The total time spent on re-evaluations is T_k. We seek a bound on T_k.

---

[4]The *WeightedExpansion* algorithm is slightly non-deterministic in that it does not specify how to break ties (line [20] in Figure 7.8) when several vertices have the minimum *leastPathCost*. Consequently, $\overline{\text{WorkSet}}$ is not precisely defined. For this proof we can use the final value of WorkSet computed by a deterministic version of *WeightedExpansion* that adds all vertices with minimum *leastPathCost* to WorkSet when there is tie.

---

procedure ImprovedWeightedExpansion (G, U)
declare
      G : an annotated circuit
      U : a set of vertices in G
      WorkSet, *ApparentlyAffected*, *PotentiallyAffected*: sets of vertices
      u, v, w: vertices
      WorkSetCost, totalCost: integer
preconditions
      Every vertex in $V(G)-U$ is consistent and $v.leastPathCost$ is ∞ for every vertex v
begin
[1]   /* Initialize WorkSet */
[2]   *totalCost* := 0
[3]   *WorkSetCost* := 0
[4]   WorkSet := U
[5]   **for** every vertex $u \in U$ **do**
[6] $u.originalValue$:= $u.value$
[7] $u.leastPathCost$:= $u.cost$
[8]   *WorkSetCost* := *WorkSetCost* + $u.cost$
[9]   **for** every vertex $w \in Succ(u)$ **do**
[10] $w.leastPathCost$:= min $(w.cost+u.leastPathCost, w.leastPathCost)$
[11]   **od**
[12]   **od**
[13]   **loop**
[14]   /* Evaluate WorkSet */
[15]   **for** every $v \in$ WorkSet in *relative topological order* **do** recompute $v.value$ **od**
[16]   *totalCost* := *totalCost* + *WorkSetCost*
[17]   /* Test for termination */
[18]   *ApparentlyAffected* := $\{ v \in$ WorkSet : $v.value \neq v.originalValue \}$
[19]   *PotentiallyAffected* := *Succ* (*ApparentlyAffected*) - WorkSet
[20]   **if** *PotentiallyAffected* = $\varnothing$ **then** exit loop **fi**
[21]   /* Iteratively expand WorkSet */
[22]   Choose a vertex v from *Succ* (*WorkSet*)–*WorkSet* for which $v.leastPathCost$ is minimum
[23]   **repeat**
[24]   Insert v into WorkSet
[25] $v.originalValue$:= $v.value$
[26]   *WorkSetCost* := *WorkSetCost* + $v.cost$
[27]   **for** every vertex $w \in Succ(v)$ **do**
[28] $w.leastPathCost$:= min $(w.cost+v.leastPathCost, w.leastPathCost)$
[29]   **od**
[30]   **if** *Succ* (*WorkSet*)–*WorkSet* = $\varnothing$ **then** exit inner loop **fi**
[31]   Choose a vertex v from *Succ* (*WorkSet*)–*WorkSet* for which $v.leastPathCost$ is minimum
[32]   **until** *WorkSetCost* + $v.cost$ > *totalCost*
[33]   **end loop**
[34]   **for** every vertex $v \in$ (WorkSet $\cup$ *Succ* (WorkSet)) **do** $v.leastPathCost$:= ∞ **od**
end
postconditions
      Every vertex in G is consistent $v.leastPathCost$ is ∞ for every vertex v

---

Figure 7.9. An improved incremental algorithm for the weighted circuit value annotation problem.

If k is 1, then the algorithm takes only time $O(Cost(U))$, i.e. $O(C_\delta)$. For $k \geq 2$, $T_k = W_{k-1} + T_{k-1}$. We obtain bounds on each of W_{k-1} and T_{k-1} below.

If k is 2, then $T_{k-1} \leq Cost(U) \leq Cost(\overline{\text{WorkSet}})$. Now consider $k \geq 3$. Consider the $k-2$-th iteration of the outer loop. Now $Succ(\text{WorkSet}_{k-2}) - \text{WorkSet}_{k-2}$ must be nonempty, since the algorithm makes at least two more iterations. Let $\overline{v}$ be the vertex in $Succ(\text{WorkSet}_{k-2}) - \text{WorkSet}_{k-2}$ for which $\overline{v}.leastPathCost$ is minimum. The termination condition for the inner loop (in the $k-2$-th iteration of the outer loop) implies that $W_{k-2} + \overline{v}.cost > T_{k-2}$. Since the algorithm takes two more iterations, $\text{WorkSet}_{k-2} \cup \{\overline{v}\} \subseteq \overline{\text{WorkSet}}$. Hence $T_{k-2} < W_{k-2} + \overline{v}.cost \leq Cost(\overline{\text{WorkSet}})$. Since $T_{k-1} = W_{k-2} + T_{k-2}$, it follows that $T_{k-1} \leq 2 \cdot Cost(\overline{\text{WorkSet}})$.

Now we obtain a bound on W_{k-1}. Consider the $k-1$-th iteration of the outer loop.

Case 1. Assume that $W_{k-1} \leq T_{k-1}$. Then, $T_k = T_{k-1} + W_{k-1} \leq 2 \cdot T_{k-1} \leq 4 \cdot Cost(\overline{\text{WorkSet}})$.

Case 2. Assume $W_{k-1} > T_{k-1}$. Consider the termination condition in line [32]. It attempts, in the i-th iteration of the outer loop, to ensure that $W_i \leq T_i$, by avoiding adding any vertex v to WorkSet that will make $W_i > T_i$. However, the repeat loop (lines [23-32]) will execute at least once, and the only way the condition $W_i \leq T_i$ can fail is if the repeat loop executes exactly once (during the i-th iteration of the outer loop) and the very first vertex added to WorkSet makes its cost greater than T_i. Hence, if $W_{k-1} > T_{k-1}$, then it must be the case that $\text{WorkSet}_{k-1} = \text{WorkSet}_{k-2} \cup \{\overline{v}\}$, where $\overline{v}$, as above, is the vertex in $Succ(\text{WorkSet}_{k-2}) - \text{WorkSet}_{k-2}$ for which $\overline{v}.leastPathCost$ is minimum. Thus, $W_{k-1} = Cost(\overline{\text{WorkSet}}_{k-1}) \leq Cost(\overline{\text{WorkSet}})$.

It follows that $T_k = W_{k-1} + T_{k-1} \leq 3 \cdot Cost(\overline{\text{WorkSet}})$. $\square$

7.6. The Empirical Boundedness of Incremental Algorithms

In this section, we present the results of a study in which we explored the use of the parameter $\|\delta\|$ in the empirical evaluation of incremental algorithms. Earlier, in Section 2.3.2, we discussed the importance of the complexity parameter used in experimental evaluation of incremental algorithms, and suggested that the parameter $\|\delta\|$ could be useful in such experimental studies too. There are a couple of reasons for this approach. One is that even though worst-case analysis might indicate that an algorithm was unbounded, it could be the case that on the inputs that arise in practice the algorithm does behave like a bounded algorithm, and this is worth knowing. The second reason is that if an algorithm behaves in a bounded fashion on inputs that arise in practice, its complexity is likely to correlate more to $\|\delta\|$ than to the input size. In such cases, studying the algorithm's complexity as a function of $\|\delta\|$ is likely to shed light on the algorithm's performance better.

Our study involved three incremental algorithms for the circuit value annotation problem: *RBF_Expansion*, a variant of *RBF_Expansion*, and the priority-ordering based update algorithm due to Alpern et al. [Alp90], which we will call *PO_Update*. All these updating algorithms were implemented in the Synthesizer Generator, a system for automating the construction of customized editors for particular languages

[Rep88b, Rep88]. The Synthesizer Generator was then used to generate a Pascal editor that does incremental static-semantic checking of the program being edited using the above-mentioned incremental algorithms. This enabled us to test the performance of the updating algorithms easily, by performing an extensive sequence of editing operations on Pascal programs of varying size.

The circuit for which incremental updating is performed is in this case the dependence graph of the attribute instances of the program being edited. Consequently, these circuits have some structure to them, and the performance results presented in this section need not necessarily carry over to the problem of updating the annotations of arbitrary circuits. However, the results presented here are promising.

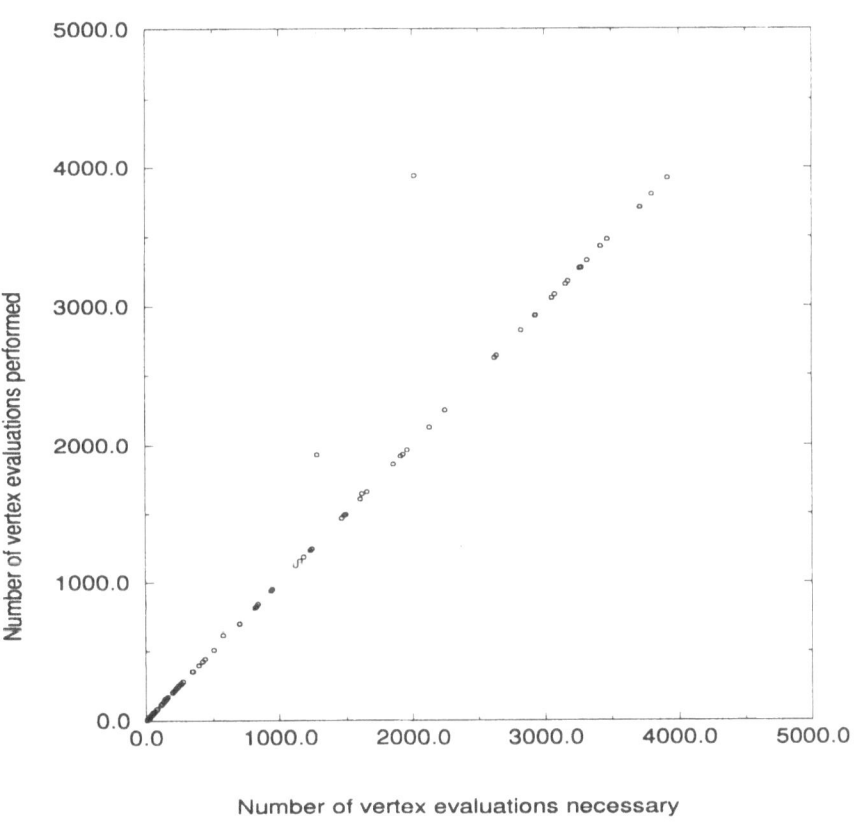

Number of vertex evaluations necessary

Figure 7.10. The above graph shows the number of vertex evaluations that Improved_RBF_Expansion performed as a function of the minimum number of evaluations any algorithm must perform. Each point in the graph represents an editing operation. Observe that most of the points fall very close to the $y = x$ line.

In reporting the results of our experiments, we do not present actual running times for these incremental algorithms. Instead, we compare the number of operations performed by the updating algorithms as a function of $\| \delta \|$, which is a lower bound on the number of operations any updating algorithm has to perform. These results enable us to identify empirically the "boundedness behavior" of these two incremental algorithms. One of the reasons actual timing results have not been presented is that they depend both on the machine and the implementation details. The figures presented in this section provide a more machine-independent and implementation-independent view of the situation.

The Performance of RBF_Expansion and Improved_RBF_Expansion

The algorithm we implemented and tested is a variant of *RBF_Expansion* that incorporates the improvements outlined in Remark 7.2—we will refer to this as the *Improved_RBF_Expansion* algorithm. One of the chief characteristics that distinguishes this algorithm from other algorithms for this problem, such as the *PO_Update* algorithm discussed next, is that it does not maintain any information relating to the topological ordering of the circuit across invocations. It begins the updating essentially with no information about transitive dependences between vertices. Hence, the algorithm can end up evaluating vertices in the wrong order. As a consequence, the algorithm can perform more vertex evaluations than necessary, both because some vertices can be evaluated multiple times and because some vertices that are not potentially affected might be evaluated.

The graph in Figure 7.10 plots the number of vertex evaluations the algorithm performed against the minimum number of vertex evaluations any algorithm must perform (that is, $|\text{CHANGED} \cup Succ\,(\text{AFFECTED})|$). (See the discussion on the minimum number of evaluations necessary in Chapter 3.) As the graph shows, this particular algorithm often performs close to the minimum number of evaluations. The algorithm performed exactly the minimum number of evaluations necessary in approximately 70% of all the updates, and performed less than 1% extra evaluations in approximately 90% of all the updates. In some rare worst cases it performed approximately two to three times the minimum number of vertex evaluations necessary. Since the time complexity of the algorithm is linear in the number of vertex evaluations it performs (in bounded degree graphs, such as the dependence graphs that arise in the attribute updating problem), obviously this algorithm's performance compares reasonably with an asymptotically optimal algorithm's performance.[5]

---

[5]Similar results were reported by Hoover for his incremental algorithm for the same problem, which utilized an approximate topological sort ordering to update the circuit [Hoo86a]. However, not too much weight should be attached to the exact figures quoted above. The results suggest that our algorithm performs close to the minimum number of evaluations necessary for the more common editing operations, but there are worst-case editing operations where the algorithm may perform approximately twice or thrice the number of evaluations necessary. The percentages we list above will obviously depend on the mix of editing operations used in the experimentation. We tried to uniformly create editing operations of different categories.

The graph in Figure 7.11 similarly plots the number of vertex evaluations the algorithm would have performed if the improvements suggested in Remark 7.2 had not been incorporated. As the graph shows, the algorithm would have performed very poorly, with a complexity that appears to be quadratic in $\| \delta \|$.

The Performance of the Priority-Ordering Based Update Algorithm
We first briefly explain the idea behind the algorithm *PO_Update* outlined in [Alp90]. *PO_Update* uses a prioritization of the dag as auxiliary information for the circuit value annotation problem. Recall that a prioritization of a dag associates each vertex

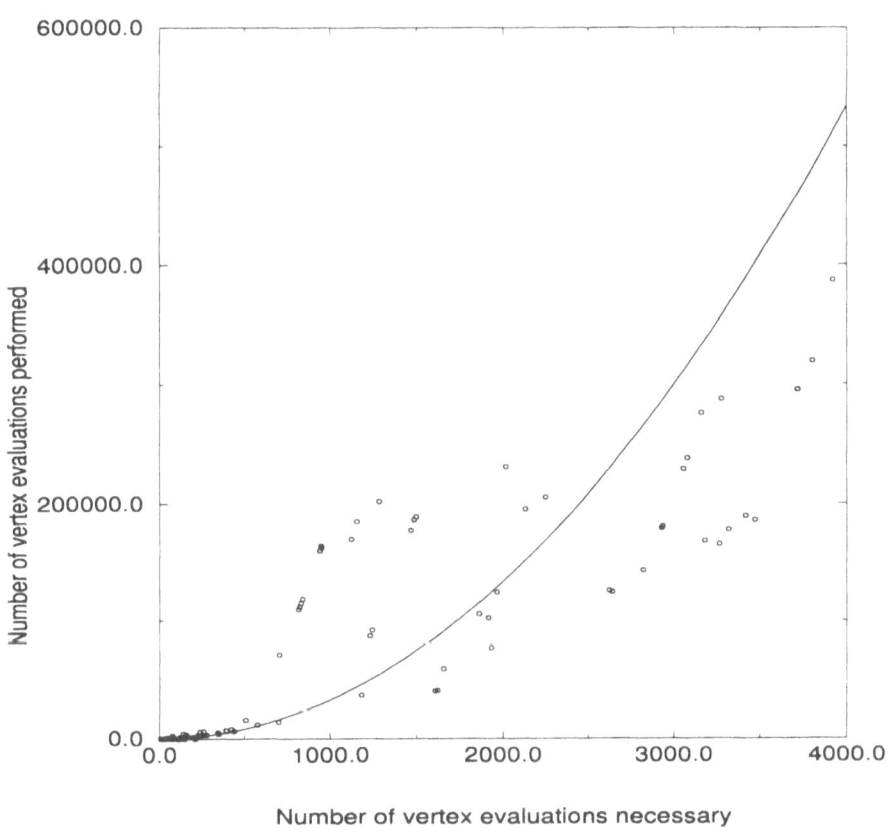

Figure 7.11. The above graph shows the number of vertex evaluations that the straight forward implementation of RBF_Expansion perform as a function of the minimum number of evaluations any algorithm must perform. Observe that the scales of the two axis are not the same. The above graph also shows a plot of the function $y = x^2 / 30$ for comparison. As the graph shows, the behavior of this simple implementation of RBF_Expansion is non-linear in $\| \delta \|$, and is approximately quadtratic in $\| \delta \|$.

in the dag with a priority, which is a value drawn from a totally ordered set, such that if there is a path from a vertex u to a vertex v, then u's priority is less than v's priority. After a change to the circuit, *PO_Update* first updates the priorities of vertices to obtain a correct prioritization of the current circuit. Then, vertex re-evaluations are scheduled (via a worklist algorithm that uses a priority queue for the worklist). This algorithm runs in time

$$\| \delta_{PriorityOrdering} \|^2 \log \| \delta_{PriorityOrdering} \| + \| \delta_{CircuitValue} \| \log \| \delta_{CircuitValue} \|$$

where $\| \delta_{PriorityOrdering} \|$ denotes the size of the change in the input and the auxiliary information, while $\| \delta_{CircuitValue} \|$ denotes the size of the change in the input and the output of the circuit value annotation problem. Because the quantity $\| \delta_{PriorityOrdering} \|$ is not bounded by any function of $\| \delta_{CircuitValue} \|$, this algorithm for the incremental circuit-value problem is unbounded. In other words, we can construct a sequence of circuits and associated changes in the circuit such that there is no bound on the value of $\| \delta_{PriorityOrdering} \|$ in these cases, even though $\| \delta_{CircuitValue} \|$ is bounded by some constant in all these cases.

Unlike *RBF_Expansion*, this algorithm evaluates the vertices in the right order. Hence, it does not have the overhead of extra vertex evaluations. However, it does have other overheads: the overhead of maintaining auxiliary information, and a logarithmic overhead due to the use of a heap to evaluate vertices in the correct order.

We measured the values of both $\| \delta_{PriorityOrdering} \|$ and $\| \delta_{CircuitValue} \|$ over a number of editing sessions, and the results are shown in Figure 7.12. These results show that, at least in the case of a Pascal editor, this algorithm behaves effectively like a bounded incremental algorithm. The reason is that, as the associated graphs show, $\| \delta_{PriorityOrdering} \|$ is invariably less than or equal to $\| \delta_{CircuitValue} \|$. In fact, while it was possible to generate editing operations that produced larger and larger values of $\| \delta_{CircuitValue} \|$ for larger and larger programs, $\| \delta_{PriorityOrdering} \|$ seemed to be intrinsically bounded by some constant.

7.7. Some Remarks

In this chapter we studied several, closely related, exponentially bounded incremental algorithms for the circuit value annotation problem. We also discussed the performance of some incremental algorithms in a real world example. These empirical results show that in the case of unbounded algorithms and exponentially bounded algorithms one can potentially run into the standard drawbacks of worst-case analysis—the behavior of incremental algorithms might be far better on inputs that arise in practice than for worst-case instances. These results also highlight another drawback of asymptotic worst-case analysis: simple improvements, which do not change the asymptotic worst-case behavior of an algorithm, may greatly improve the performance of the algorithm on inputs that arise in practice. Our experimental findings must be tempered by the fact the algorithms were tested against a restricted class of inputs, namely for circuits and modifications that arise in a Pascal editor. There is no guarantee that the algorithms will perform similarly in the case of circuits and modifications that might arise in a more general situation.

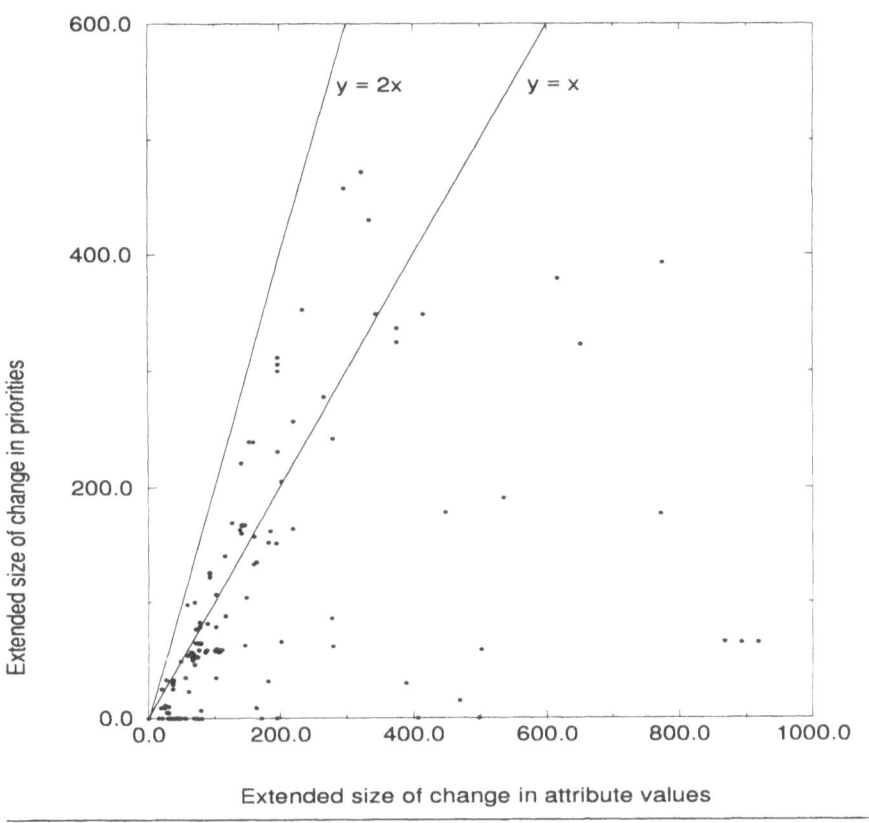

Figure 7.12. The above graph is a plot of $\|\delta\|$ for the auxiliary information (that is, priorities) versus $\|\delta\|$ for the actual attribute updating problem in the case of a Pascal editor. The graph also shows the $y = x$ line and $y = 2x$ line. It can be seen that the size of the change in the priorities tends to be bounded by the size of the change in the attribute values.

The existence of several variants of the iterative evaluate-and-expand strategy with incomparable performance characteristics also suggests the possibility of choosing the updating algorithm for each application based on characteristics of the circuits that arise in that particular application.

> *Reason possesses insights only into that which reason itself constructs according to its own plan, and though reason may take the lead with its own proposals, it must then by experiment elicit from nature the wisdom of these proposals. There is a time for theory and a time to decide the disposition of that theory by nature's behavior.*
>
> —Morris Kline, *Mathematics: The loss of certainity*

Chapter 8
Inherently Unbounded Incremental Computation Problems

> *The situation, not infrequent in mathematical research, is this: A theorem has been already formulated but we have to give a more precise meaning to the terms in which it is formulated in order to render it strictly correct.*
> —G. Polya, *Mathematics and Plausible Reasoning, Volume 1*

> *... proofs, even though they may not prove, certainly do help improve our conjecture.*
> —I. Lakatos, *Proofs and Refutations: The Logic of Mathematical Discovery*

8.1. Introduction

In this chapter we establish lower bounds for some incremental computation problems. In particular, we show in Section 8.3 that the single-source graph reachability problem is unbounded. We establish in Section 8.4 that various other problems, such as dataflow analysis problems of various kinds and algebraic path problems, are unbounded by reducing the incremental version of the reachability problem to other incremental problems.

Lower bounds are usually established with respect to a model of computation. The lower bounds in this chapter are established with respect to a model of computation called *sparsely-aliasing pointer machine*. This is a considerably more powerful model than the model of *locally persistent algorithms* [Alp90], which has been previously used for lower bound proofs in incremental computation, though it is, strictly speaking, not a generalization of the model of locally persistent algorithms. The lower bound proof can, however, be easily adapted for locally persistent algorithms and for various extensions of that model. These different models of computation are discussed in Section 8.2.

8.2. The Model of Computation

8.2.1. Locally Persistent Algorithms

The class *LP* of *locally persistent* algorithms was introduced by Alpern *et al.* in [Alp90]. What follows is their description of this class of algorithms, paraphrased to be applicable to general graph problems: A *locally persistent* algorithm may make use of a block of storage for each vertex of the graph.[1] The storage block for vertex u

---

[1] This may be directly generalized to permit storage blocks to be associated with edges too.

will include pointers to (the blocks of storage for) the predecessor and successor vertices of u. The storage block for u will contain the output value for u. The block may also contain an arbitrary amount of auxiliary information, but no auxiliary pointers (to vertices, *i.e.*, their storage blocks). No global auxiliary information is maintained in between successive modifications to the graph: whatever information persists between calls on the algorithm is distributed among the storage blocks for the vertices. An input change is represented by a pointer to the vertex or edge modified. A locally persistent algorithm begins with the representation of a change and follows pointers. The choice of which pointer to follow next may depend (in any deterministic way) on the information at the storage blocks visited so far. For example, a locally persistent algorithm may make use of worklists or queues of successors of vertices that have already been visited. The auxiliary information at a visited storage block may be updated (again in any way that depends deterministically on the information at the visited storage blocks).

In summary, an *LP* algorithm has two chief characteristics. First, any auxiliary information used by the algorithm is associated with an edge or a vertex of the graph—no information is maintained globally. Second, the algorithm starts an update from the vertices or edges that have been modified and traverses the graph using only the edges of the graph. In essence, the auxiliary information at a vertex or edge cannot be used to access non-adjacent vertices and edges.

Alpern et al. utilized this model of computation in establishing a lower bound for the incremental circuit value problem. They showed that any bounded *LP* algorithm for the circuit value problem must take time at least $\Omega(2^{||\delta||})$ to process a change δ. Subsequently, Ramalingam and Reps [Ram91] showed that the dynamic single-source reachability problem (the SS-REACHABILITY problem) and some other graph problems have no bounded *LP* algorithm. A similar approach was subsequently used by Berman [Ber92] to show that other problems, such as the connected components problem, are unbounded with respect to the class *LP*.

However, the model of *LP* algorithms is quite a weak model of computation. All of the above mentioned lower bound proofs rely heavily on the fact that no information may be stored globally in the case of *LP* algorithms. But if one considers a problem like reachability, it is unreasonable to restrict an incremental algorithm from maintaining a pointer to the source vertex. Once we allow the incremental algorithm to maintain such a pointer, any auxiliary information maintained at the storage block for the source vertex is in effect globally available information, and all the above lower bounds fail!

Before we define a suitable generalization of the class *LP* we consider the cost of performing elementary operations.

8.2.2. The Cost of Elementary Operations

In the *logarithmic cost measure*, each basic operation is assigned a cost proportional to the number of bits needed to represent the operands [Tar83]. Note that under this cost measure practically any incremental algorithm will be an unbounded algorithm, since most operands being manipulated will be $O(\log n)$ bits long, and, hence, any basic operation in the machine will have a cost of $O(\log n)$.

In analyzing algorithms it is much more common to use *uniform cost measure*, under which each basic operation is assigned a unit cost. However, if basic operations are allowed to manipulate numbers of arbitrary size, this measure can be misleading. So, it is usual to restrict the sizes of operands in basic operations to be $O(\log n)$ bits long, where n is a measure of the input size. We will use this variant of uniform cost measure in which each unit cost operation is allowed to manipulate only $O(\log n)$ bit quantities.

If there is no restriction on the size of the operands of elementary operations, one can effectively perform a number of basic operations by performing a single operation on a huge operand, and obtain algorithms with a misleadingly low cost. We now show that the version of the dynamic single-source reachability problem in which the set of vertices is static *does* have a bounded incremental algorithm in the absence of restrictions of the above kind.

Let us consider graphs over a given (static) set of vertices V. Let n be the number of vertices. There are n^2 possible edges in any graph over V. Consequently, there are 2^{n^2} possible graph instances on the given set of vertices V, since each edge can be present or absent, independent of the other edges. Consider a finite state automaton whose set of states is given by the set of 2^{n^2} possible graphs. If a graph G_2 can be obtained from a graph G_1 by the insertion or deletion of an edge then there is a transition from G_1 to G_2 in the FSA labelled with this modification. The transition is also associated with the set of vertices whose reachability status changes as a result of this modification.

Note that the above FSA can be constructed initially, as part of the preprocessing step, once the size of the input graph is known. Once the FSA has been constructed, processing a modification is simple. The current state describes the current graph. Just looking up the transition table using both the current state and the input modification as an index gives us both the new state and the list of vertices whose reachability status must be changed. This effectively means that the update can be done in $O(\|\delta\|)$ time.

Such an algorithm is unsatisfactory because of the enormous time required for the preprocessing step and the enormous space requirements for storing the transition table of the FSA. This objection, by itself, does not argue that the above algorithm is an unbounded one, since the definition of boundedness places no restriction on the amount of auxiliary storage that may be used or on the amount of time that can be spent in preprocessing. However, realistically speaking, the above algorithm is not a bounded one: with 2^{n^2} possible states, one needs n^2 bits to represent a state, and it is not reasonable to count operations with the state, such as assigning a new value to the state, as constant time operations.

In a sense, the above argument captures the essence of our lower bound proof that reachability is an unbounded problem. It is hard to imagine an incremental algorithm that would do better than the above algorithm *during* the updating step, since the above algorithm does most of the work during the preprocessing step, and does the minimal amount of work necessary during the update step. So, what more do we need to do to establish that reachability is an unbounded problem?

There is a simple point we need to address concerning the above argument. It is possible that an algorithm that uses a minimized form of the above FSA might do better. Unfortunately, the minimized FSA will not be much smaller than the above FSA. Since self loops and edges directed *to* the source vertex do not affect the reachability status of any of the vertices, the state need not contain information about the presence or absence of the above edges. But all of the remaining $(n-1)(n-2)$ edges are important, and the state needs to encode the presence or absence of these edges.

What we *need* to establish in the lower bound proof is that there is no clever enough representation of the state that allows the updating algorithm to efficiently change the state by examining and changing only a "small" number of bits in the representation of the state.

8.2.3. Sparsely Aliasing Algorithms

As explained in the previous section, the idea is to show that no clever encoding of the "state" exists that allows the updating algorithm to process "trivial" input changes in constant time. Given the way we measure the cost of elementary operations, the algorithm can examine and manipulate or modify only $O(\log n)$ bits of information in a constant number of steps. We want to show that this is not enough. To establish this, we formalize the way the updating algorithm accesses and updates its state as below.

By "state" we mean the information the incremental algorithm retains across different calls to the updating algorithm. This includes a description of the current graph, in whatever form the algorithm stores it, and any auxiliary information the algorithm might choose to maintain, and the output information.

We assume that the state is represented as in the pointer machine model [Tar83]. The memory of a pointer machine model is an expandable collection of nodes or blocks. Each block is divided into a fixed number of fields, each of which holds either an atomic value (such as an integer or boolean value) or a pointer to a block. The value stored in a field of a block can be retrieved only if a pointer to that block is available.

Thus, there are a bounded number of field selectors, which we will denote f_1, $f_2, \ldots, f_m$. If p is a pointer to a block, then $f_i(p)$ denotes the value in the i-th field of the block. Every piece of information stored in the memory must be retrieved via a sequence of pointer dereferencing. Though the memory can contain an unbounded number of blocks, there must be a bounded number of "base" pointers which are used to access any piece of information. Consider standard algorithms which create, modify, and use dynamic data structures: the data structures might grow unbounded, but all the information in the data structures is accessible via a bounded number of base or root pointers. (Information not so accessible is really "garbage" that cannot be used.)

In the case of the reachability problem, a pointer to the source vertex and a pointer to a linked list of all vertices in the graph are examples of such base pointers.

In the case of an incremental or dynamic algorithm, the base pointers are of two types: "variables" whose values are retained across different invocations of the incremental algorithm (these are like the global variables and static local variables of C), and "input arguments" which are passed to the updating algorithm. Thus, any incremental algorithm for reachability may maintain a bounded number of pointers $P = \{p_1, \ldots, p_r\}$. The updating algorithm is given as an argument a description of the input change, which includes pointers to the modified vertices. (We assume that the memory contains a block for each vertex in the graph, and that a vertex is represented by a pointer to the corresponding block.) See Figure 8.1.

The state of the system is, thus, the contents of the pool of blocks *and* the values stored in the set of pointers P. But we will often use the word "state" to denote just the contents of the collection of blocks.

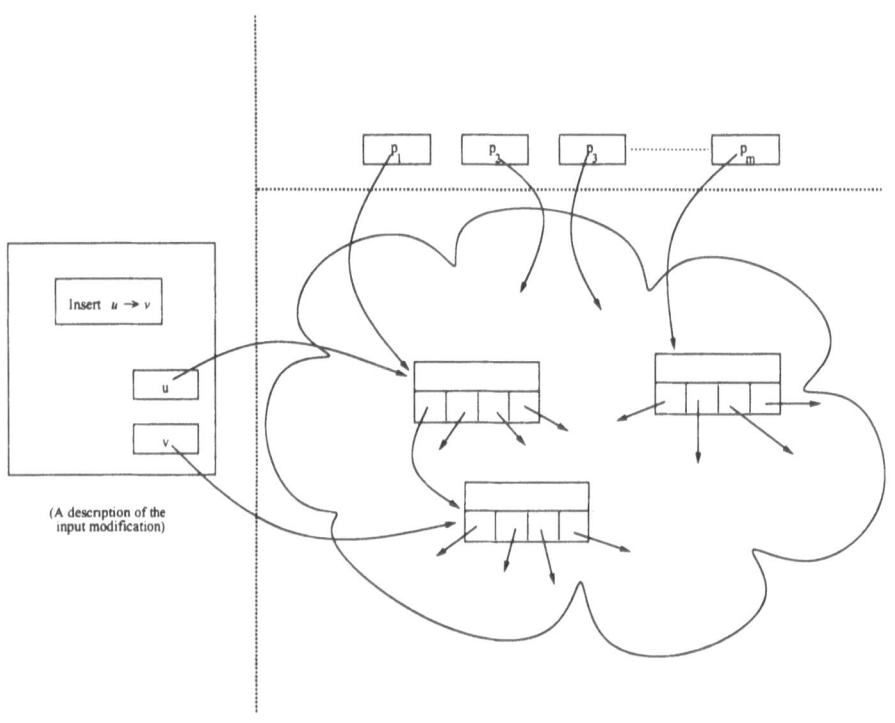

Figure 8.1. The organization of the information that an incremental algorithm retains across different invocations, and its retrieval from a set of base pointers. The state information is distributed across a potentially unbounded number of blocks, but all information used by the algorithm must be retrieved via a sequence of pointer dereferences starting from one of a bounded number of base pointers. In the case of incremental algorithms, this set of base pointers includes a set of global pointers $\{p_1, \ldots, p_m\}$ and a set of pointers to the modified vertices that is passed in as an argument to the algorithm.

The structure of a state can be described by a graph as follows. The *representation graph* of a state σ consists of one vertex for each block in the state. It contains an edge from the vertex representing a block b_i to a vertex representing a block b_j iff the block b_i has a pointer to block b_j. Assume that an input graph G and associated information is represented by a state σ. Then, since a vertex in G is represented by a block in σ, the representation graph of σ contains a vertex for each vertex in G. It may, in general, contain a number of other vertices and edges that have no correspondence to edges and vertices in G. The model of *LP* algorithms requires that the undirected version of the representation graph be, in fact, isomorphic to the undirected version of the given graph, though it does not restrict the blocks to be of fixed size.

The indegree of a vertex is the number of predecessors it has. The outdegree of a vertex is the number of successors it has. The degree of a vertex is the sum of its indegree and its outdegree. The indegree, outdegree, or degree of a graph is defined to be the maximum indegree, outdegree, or degree of any vertex in the graph. Since each block has only a bounded number of fields, the outdegree of the representation graph of a state σ is bounded by a constant. There is no such bound on the indegree of the representation graph.

We now introduce various classes of incremental graph algorithms, characterized by the "sparsity" (in terms of indegree) of the representation graphs they use. An incremental graph algorithm is said to be an *[f(n), g(n)] restricted aliasing algorithm* if in any representation σ corresponding to a graph of n vertices at most $g(n)$ vertices have indegree greater than $f(n)$. An incremental graph algorithm is said to satisfy the *sparse-aliasing* condition if it is a $[O(\log^r(n)), O(n^\alpha)]$ restricted aliasing algorithm for some constant r and some constant $\alpha < 0.5$. We call such algorithms *SA* algorithms.[2]

Let us now compare the class of *LP* algorithms with the class of *SA* algorithms. As explained above, the class of *LP* algorithms satisfy much more stringent requirements on the structure of the representation graph than the class of *SA* algorithms. Since the undirected versions of the actual input graph and the representation graph need to be isomorphic in the case of *LP* algorithms, the state σ of any *LP* algo-

---

[2]Note that if one uses various kinds of adjacency list representations of a graph G then one would need to have up to *degree*(G) pointers to the same vertex. Consequently, this definition is too restrictive in that one cannot use adjacency list representations in the case of dense graphs. A more appropriate approach would be to define an algorithm to be an $[f(n,d), g(n)]$ restricted aliasing algorithm if in any representation σ corresponding to a graph of n vertices and degree d, at most $g(n)$ vertices have indegree greater than $f(n,d)$. Then, we could define an algorithm to be an *SA* algorithm iff it is an $[f(n,d), g(n)]$ restricted aliasing algorithm for some function $f(n,d)$ that is poly-logarithmic in n for a fixed d, and a function $g(n) = O(n^\alpha)$ for some $\alpha < 0.5$. However, our lower bound proof will involve only graphs of degree at most three. Hence, we are interested only in bounded degree graphs, and the above definition suffices.

rithm corresponding to an input graph G satisfies *indegree*(σ) $\leq$ *degree*(G). Consequently, any *LP* algorithm is a [d, 0] restricted aliasing algorithm, although the blocks in *LP* algorithms are not required to be of bounded size. Thus, *LP* algorithms would trivially be *SA* algorithms, but for the fact that *LP* algorithms place no restriction on how much information can be stored in a single block, or on the way information may be stored or organized *within* a block corresponding a vertex. The definition of local persistence could afford to ignore the organization of information within a block primarily because its other requirements were stringent enough to establish the required lower bounds.

Let us now see how the class of *SA* algorithms includes many algorithms that are not *LP* algorithms. The definition of *SA* algorithms removes two of the biggest restrictions in the definition of *LP* algorithms. The first restriction relates to the use of "global state". We will refer to the part of the state σ accessible from the set P of pointers as the "global" state. The significance of global state is that it is accessible no matter what the input modification is. This is in contrast to the model of locally persistent algorithms, where no such global information is available. In a *LP* algorithm every piece of information is stored in a block associated with a vertex, and one needs a pointer to a vertex in order to access the information associated with that vertex. This, in turn, implies the following. Consider a sequence of modifications δ_1 and δ_2 to a graph G. The only way the updating done after δ_1 can affect the way δ_2 is processed subsequently is by modifying the information associated with a vertex that is subsequently examined during the processing of δ_2. In contrast, with global information, the processing of δ_1 can change some information in the global state and directly affect the way δ_2 is subsequently processed.

The second restriction removed in the definition of *SA* algorithms relates to the use of "non-local pointers". The block representing a vertex u may store pointers to blocks representing vertices that are non-adjacent to u in the graph, as long as the sparse-aliasing conditions are met.

Yet there remain some restrictions on the model of computation. The biggest restriction is the sparse-aliasing condition. The second restriction, which is common to all pointer machine models, is that address arithmetic is not allowed. We will discuss these restrictions again in the conclusion section. Though the sparse-aliasing restriction is somewhat artificial, the model of computation is a great improvement on the previous model of *LP* algorithms. Further, the definition of the model of *SA* algorithms reflects the lower bound proof, and suggests some features that an incremental algorithm *must* possess if it is to defeat the lower bound argument. Thus, understanding *SA* algorithms could be useful in designing bounded incremental non-*SA* algorithms for problems such as reachability and dataflow analysis, though the author suspects that the unboundedness results hold with respect to even stronger models of computation.

8.3. The Unboundedness of Reachability

In this section we show that the single-source reachability problem has no bounded *SA* algorithm. We will consider only bounded degree graphs in which the set of vertices remains static. We will restrict our attention to "almost-unit" changes: input modifications that insert at most one edge and delete at most one edge. (We can recast the proof so that it makes use of only unit changes: input modifications that either insert a single edge or delete a single edge.) Since our proof establishes that even this restricted dynamic reachability problem is unbounded, it follows that the more general, fully dynamic, reachability problem must be unbounded as well.

An input modification is said to be redundant if it does not change the reachability status of any vertex in the graph. Since we consider only bounded degree graphs and almost-unit changes, there exists a constant c such that $\|\delta\| \le c$ for any redundant change δ. In particular, any bounded incremental algorithm for reachability should be able to process redundant input changes in constant time.

We will consider a simpler form of the dynamic reachability problem below. In this simpler problem, the algorithm has to respond to every input modification by classifying it as a redundant modification or irredundant modification. A *bounded redundancy checking algorithm* is an algorithm that responds to redundant input modifications with an "yes" in constant time, and one that responds to input modifications that are not redundant with a "no", taking any amount of time for the processing. Obviously any bounded incremental algorithm for reachability can be used as a bounded redundancy checking algorithm. We now establish

Theorem 8.1. *The single-source reachability redundancy checking problem has no bounded SA algorithm.*

The proof is by contradiction. We first establish a sequence of propositions, which will lead us to the contradiction.

Let us assume that there is a bounded *SA* algorithm $\mathcal{A}$ for checking the redundancy of input modifications. Let k be a constant such that $\mathcal{A}$ performs no more than k elementary operations in processing any redundant input modification. Assume that the updating algorithm is passed a set U of vertices as input argument. (Thus, U is the set of modified vertices.) Then, the updating algorithm *can examine and modify only the part of the state that is reachable from the set $U \cup P$ of pointers*. Further, assume that the algorithm performs no more than k elementary operations. Then, it can examine or modify only the part of the state *that is reachable in k or less steps* from the set $U \cup P$ of pointers.

We formalize some of these concepts below. If X is a set of pointers, then $\sigma \downarrow (X, k)$, the part of the state σ accessible in k steps or less from X, is a function from terms of the form $(f_{i_1} \circ f_{i_2} \circ \cdots \circ f_{i_r})(u)$ to their values where $r \le k$, and $u \in X$. Observe that there are at most $(m+1)^k |X|$ such terms, since there are only m different fields in a block. Since the value of each term is an $O(\log n)$ bit quantity/word, the value of $\sigma \downarrow (X, k)$ can be completely specified by an $O(\log n)$ bit quantity if $|X|$ is considered to be a constant.

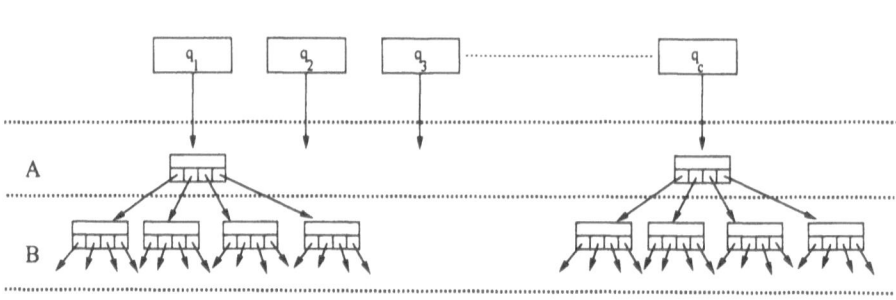

Figure 8.2. We denote the part of the state σ accessible in k steps or less from a set X of pointers by $\sigma \downarrow (X,k)$. In the above picture, A and B denote the sets of blocks accessible from a set of pointers $\{q_1, \ldots, q_c\}$ in 1 and 2 steps respectively. For a set X of cardinality bounded by some constant, and for a fixed constant k, $\sigma \downarrow (X,k)$ contains only $O(\log n)$ bits worth of information.

Note that the restriction that every field be an $O(\log n)$ bit quantity forces pointers also to be only $O(\log n)$ bits long. This, in turn, implies that the number of blocks allowed, in the state, is not really unbounded—it has to be bounded by some polynomial in n, the number of vertices. We can relax this restriction and allow pointers to be arbitrarily long, as long as the algorithm uses pointers solely for retrieving values. Thus, the bit representation of the pointer cannot be used to encode any kind of information. With this restriction, again the restricted state $\sigma \downarrow (X,k)$ can contain only $O(\log n)$ bits of useful information.

Proposition 8.2. Assume that σ_1 and σ_2 are such that $\sigma_1 \downarrow (\text{MODIFIED}_\delta \cup P,k) = \sigma_2 \downarrow (\text{MODIFIED}_\delta \cup P,k)$. Then, δ is redundant for σ_1 iff it is redundant for σ_2.

Proof. This follows from the choice of k and the fact that the updating algorithm is deterministic. Assume that δ is redundant for σ_1. Then, the updating of δ on σ_1 halts after performing no more than k operations. Each of these operations uses values retrieved only from $\sigma_1 \downarrow (\text{MODIFIED}_\delta \cup P,k)$. Since $\sigma_1 \downarrow (\text{MODIFIED}_\delta \cup P,k) = \sigma_2 \downarrow (\text{MODIFIED}_\delta \cup P,k)$, the updating algorithm must behave identically when it is processing change δ in state σ_2 too. Consequently, δ must be redundant for σ_2 also. $\square$

Our approach to establishing that SS-REACHABILITY has no bounded SA algorithm is to derive a contradiction using the above proposition; the contradiction is obtained by constructing two graphs G_1 and G_2 represented by states σ_1 and σ_2, and a modification δ such that δ is redundant for one of the graphs but not the other, even though the restricted state for δ is identical in both σ_1 and σ_2. We will first establish the result for $[O(\log^r(n)), 0]$ restricted-aliasing algorithms, and then generalize the result for any SA algorithm.

We can express the restricted state $\sigma \downarrow (\text{MODIFIED}_\delta \cup P,k)$ as the union of the restricted local-state $\sigma \downarrow (\text{MODIFIED}_\delta,k)$ and the restricted global-state $\sigma \downarrow (P,k)$.

(Note that the local-state and global-state need not be disjoint: there can be blocks which are accessible both from some modified vertex u and some global pointer p_i.) We will first establish that the $O(\log n)$ bits of information that can be stored in the restricted global-state cannot be of much use (for certain types of graphs).

Consider the graph G shown in Figure 8.3a. (Straight arrows in the figure indicate edges of the graph, while wavy lines indicate paths of the indicated length.) All vertices in this graph are reachable from the source vertex s. The graph has a long path, from v_0 to v_{t+1}, of length $O(n)$. This path is divided into $t+1$ smaller paths of length $l = O(n/t)$ each, where the i-th path goes from v_{i-1} to v_i.

Let σ denote the state of the system corresponding to input graph G. Consider the set of all graphs that can be obtained by adding zero or more cross edges to G, where a cross-edge is an edge from u_i to v_i for $1 \le i \le t$. A cross-edge $u_i \rightarrow v_i$, when inserted into the graph G, introduces an alternative path from the source vertex to vertex v_i (and all other vertices in the path from v_i to v_t).

For any set X of cross-edges, let σ_X denote the state of the system after it has processed the insertion of the edges in X (in a specific order) and let G_X denote the corresponding graph.

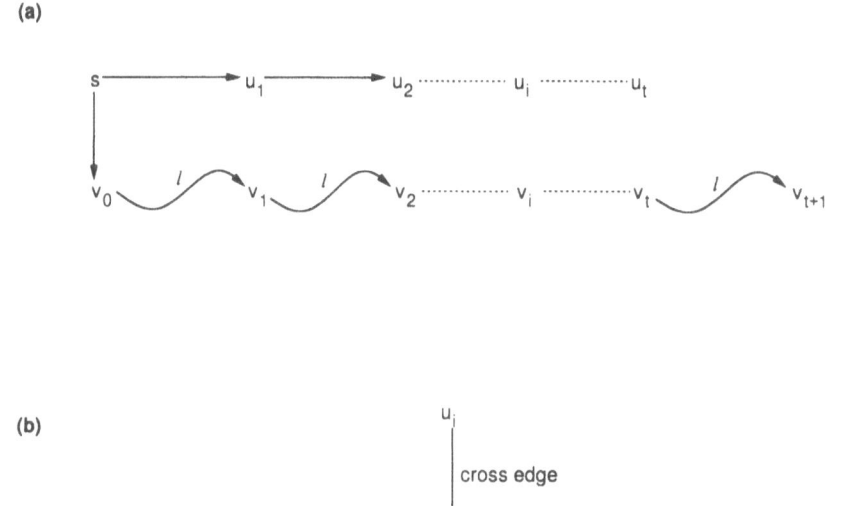

(a)

(b)

Figure 8.3. The graphs used to establish the unboundedness of reachability.

Proposition 8.3. For a suitable choice of n and t, there exist different sets of cross-edges X and Y such that $\sigma_X \downarrow (P,k) = \sigma_Y \downarrow (P,k)$.

Proof. This follows from a simple counting argument. We observed earlier that the restricted global-state contains only $c_1 \log n$ bits of information, for some constant c_1. This implies that the restricted global-state can take at most $2^{c_1 \log n}$ different values. If we choose t to be greater than $c_1 \log n$, then the restricted global-state must be identical for some σ_X and σ_Y, since there are 2^t different graphs. $\square$

We are now ready to prove our main result, Theorem 8.1, by deriving a contradiction of Proposition 8.2.

Let X and Y be two different sets of cross-edges such that $\sigma_X \downarrow (P,k) = \sigma_Y \downarrow (P,k)$. Assume, without loss of generality, that $X-Y$ is non-empty. Let $u_i \rightarrow v_i$ be a cross-edge in $X-Y$. Consider Figure 8.3b. Let P_1 denote the path from v_{i-1} to v_i, and let P_2 denote the path from v_i to v_{i+1}. We will refer to an edge $x \rightarrow y$ in path P_1 as a forward edge. We will refer to any edge from a vertex z in P_2 to a vertex in P_1 as a backedge.

Note that neither of the graphs G_X and G_Y has any backedges. Back edges can be potentially inserted into these graphs, while forward edges can be potentially deleted from these graphs.

For any edge $x \rightarrow y$ in P_1 and vertex z on P_2, let $\delta_{x,y,z}$ denote the simultaneous insertion of the back-edge $z \rightarrow y$ and the deletion of the forward edge $x \rightarrow y$. In graph G_X, this modification does not change the reachability status of any of the vertices because the cross-edge $u_i \rightarrow v_i$, the path from v_i to z and back-edge $z \rightarrow y$ combine to provide an alternative path to vertex y. In contrast, the same modification causes vertex y and some other vertices to become unreachable in the case of graph G_Y. Thus, $\delta_{x,y,z}$ is redundant for G_X but not for G_Y.

This means that when processing input modification $\delta_{x,y,z}$ the updating algorithm has to examine the state enough to determine whether the input graph is G_X or G_Y. Further, it has to determine this in no more than k steps. But we just saw in Proposition 8.3 that the updating algorithm cannot make this distinction by examining the restricted global-state.

Proposition 8.4. There exists an edge $x \rightarrow y$ in path P_1 and a vertex z in path P_2 such that $\sigma_X \downarrow (\{x,y,z\},k) = \sigma_Y \downarrow (\{x,y,z\},k)$.

Proof. Our goal is now to show that there exists some forward edge and back edge combination for which the restricted local-state is the same in both the states σ_X and σ_Y. Note that σ_X and σ_Y were obtained from σ. Further, the updating algorithm performed at most $|X| \cdot k$ elementary operations in deriving σ_X from σ, and no more than $|Y| \cdot k$ elementary operations in deriving σ_Y from σ. In particular, both σ_X and σ_Y were obtained from σ by performing no more than $O(t)$ assignments.

Let us say that a block in σ_X (or σ_Y) is directly touched if some field in that block was assigned to in deriving σ_X (or σ_Y) from σ. We say that a vertex u is touched if some block accessible from u in k steps or less was directly touched in σ_X or σ_Y.

Here is where we rely on the sparse-aliasing restriction. Note that at most $O(t)$ blocks are directly touched. Let d denote the maximum indegree of any block. Then, at most $O((d+1)^k t)$ vertices can be touched. Now t is $O(\log n)$ and d is bounded by some poly-logarithmic function. Consequently, only a poly-logarithmic number of vertices can be touched. But there are $O(n/t) = O(n/\log n)$ vertices in each of the paths P_1 and P_2. It follows immediately that, for sufficiently large values of n, there will exist a large number of edges $x \rightarrow y$ in path P_1 and vertices z in path P_2 such that $\sigma_X \downarrow (\{x,y,z\},k) = \sigma_Y \downarrow (\{x,y,z\},k)$. $\square$

Proof of Theorem 8.1. (for $[O(\log^r(n), 0]$ restricted-aliasing algorithms.)

Proof. It follows from Proposition 8.3 and Proposition 8.4 that there exist states σ_X and σ_Y, and a modification $\delta_{x,y,z}$ such that the restricted state for $\delta_{x,y,z}$ is identical in both σ_X and σ_Y, though $\delta_{x,y,z}$ is redundant for σ_X only. This gives us the desired contradiction of Proposition 8.2. It follows that there exists no bounded $[O(\log^r(n), 0]$ restricted-aliasing algorithm for checking the redundancy of an input modification for the SS-REACHABILITY problem. $\square$

We now show how the above proof can be adapted for any SA algorithm.

Consider an $[f(n), g(n)]$ restricted-aliasing algorithm. A block in the representation graph used by the algorithm is said to be a *dense* block if its indegree is greater than $f(n)$. The definition of restricted-aliasing algorithms implies that there can be at most $g(n)$ dense blocks in the representation graph. Let D denote the set of dense blocks in the representation graph (where each block is represented by a pointer to the block).

Observe that the following stronger version of Proposition 8.3 can be established: For a suitable choice of n and t, there exist different sets of cross-edges X and Y such that $\sigma_X \downarrow (P \cup D, k) = \sigma_Y \downarrow (P \cup D, k)$. The reason is that $|D|$ is bounded by $g(n)$. Consequently, $\sigma_X \downarrow (P \cup D, k)$ has only $O(g(n)\log(n))$ bits of information. Since $g(n)$ is $O(n^\alpha)$ for some $\alpha < 0.5$, we can easily choose t to be greater than $g(n)\log(n)$, and the result follows.

The previous proof of Proposition 8.4 assumed that the representation graph had no dense blocks. Once we have the above stronger version of Proposition 8.3, however, we no longer need this assumption. The same proof holds once we modify the notion of a "touched" vertex as follows: We say that a vertex u is touched if some block b accessible from u in k steps or less was directly touched in σ_X or σ_Y, *subject to the additional constraint that the access path from u to b not pass through a dense block.* Only $O(t \log^s(n))$ vertices can be touched, where s is some constant, and there are $O(n/t)$ vertices in each of the paths P_1 and P_2. Since t is $\Theta(g(n)\log(n))$, Proposition 8.4 follows.

It follows, then, from Proposition 8.3 and Proposition 8.4 that the dynamic single-source reachability problem has no bounded SA algorithm.

8.4. Other Unbounded Problems

In this section we show that various path problems in graphs and various dataflow analysis problems are all unbounded by reducing the reachability problem to these problems. The reductions utilize a "homomorphic embedding" of the reachability problem into these other problems.

8.4.1. Unbounded Path Problems

We now show that graph problems described using the closed-semiring formulation are all unbounded with respect to the *SA* model of computation. These problems were discussed in detail in Chapter 5. We review the definition of these problems below.

Definition 8.5. A *closed semiring* is a system $(S, \oplus, \otimes, \bar{0}, \bar{1})$ consisting of a set S, two binary operations $\oplus$ and $\otimes$ on S, and two elements $\bar{0}$ and $\bar{1}$ of S, satisfying the following axioms:

(1) $(S, \oplus, \bar{0})$ is a meet-semilattice with greatest element $\bar{0}$. (Thus, $\oplus$ is a commutative, associative, idempotent operator with identity element $\bar{0}$. The meet operator will also be referred to as the summary operator.) Further, the meet (summary) of any countably infinite set of elements $\{ a_i \mid i \in N \}$ exists and will be denoted by $\underset{i \in N}{\oplus} a_i$.

(2) $(S, \otimes, \bar{1})$ is a monoid. (Thus, $\otimes$ is an associative operator with identity $\bar{1}$.)

(3) $\otimes$ distributes over finite and countably infinite meets: $(\underset{i}{\oplus} a_i) \otimes (\underset{j}{\oplus} b_j) = \underset{i,j}{\oplus} (a_i \otimes b_j)$.

(4) $a \otimes \bar{0} = \bar{0}$.

A unary operator $*$, called *closure*, of a closed semiring $(S, \oplus, \otimes, \bar{0}, \bar{1})$ is defined as follows:

$$a^* =_{def} \overset{\infty}{\underset{i=0}{\oplus}} a^i$$

where $a^0 = \bar{1}$ and $a^{i+1} = a^i \otimes a$.

Different path problems in directed graphs are captured by different closed semirings. An instance of a given path problem involves a directed graph $G = (V, E)$ and an edge-labeling function that associates a value from S with each $e \in E$.

Consider a directed graph G, and a label function l that maps each edge of G to an element of the set S. The function l can be extended to map paths in G to elements of S as follows. The *label* of a path $p = [e_1, e_2, \cdots, e_n]$ is defined by $l(p) = l(e_1) \otimes l(e_2) \otimes \cdots \otimes l(e_n)$. If v, w are two vertices in the graph, then $C(v,w)$ is defined to be the meet (summary) over all paths p from v to w of $l(p)$:

$$C(v,w) = \underset{v \to_p w}{\oplus} l(p).$$

The closed-semiring framework for path problems captures both "all-pairs" problems and "single-source" problems. In an all-pairs problem, the goal is to compute $C(v,w)$ for all pairs of vertices $v, w \in V(G)$. In a single-source problem, the goal is to compute only the values $C(s,w)$ where s is the distinguished source vertex.

In all these problems, (unit-time) operations implementing the operators $\oplus$, $\otimes$, and $*$ are assumed to be available. More formally, let $S = (S, \oplus, \otimes, \bar{0}, \bar{1})$ be a specific closed semiring. The SS-S problem is defined as follows.

Definition 5.4. Given a directed graph $G = (V, E)$, a vertex s in V, and an edge-labeling function $l : E \rightarrow S$, the *SS-S problem* is to compute $C(s, w)$ for every vertex w in V. We say that (G, s, l) is an *instance* of the SS-S problem.

In the dynamic version of the SS-S problem that we consider, the source vertex s is assumed to be fixed.

For example, let S be the closed-semiring $(\mathcal{R}^{\geq 0} \cup \{\infty\}, \min, +, \infty, 0)$. Then, SS-$S$ is nothing other than the single-source shortest-path problem with non-negative edge lengths.

In this section we show that for any closed semiring S, the SS-S problem is unbounded. We first show that the SS-S problem is "at least as difficult as" the SS-REACHABILITY problem, even for incremental algorithms, by "reducing" the SS-REACHABILITY problem to the SS-S problem, and conclude that the SS-S problem is unbounded.

However, some caution needs to be exercised in making inferences about the unboundedness of a problem via a reduction argument. If a problem P is unbounded and can be reduced to a problem Q in the conventional sense, it does not necessarily follow that the problem Q is unbounded. For instance, consider any unbounded problem P of computing some value $S(u)$ for each vertex u of the graph. Consider the (intuitively) "more difficult" problem Q of computing $S(u)$ *and* $T(u)$ for each vertex u of the graph, where $T(u)$ is defined such that it changes whenever the input changes. For example, let $T(u)$ be the sum of the number of vertices and the number of edges in the graph. If each input change consists of the addition or deletion of a vertex or an edge, then by definition, whenever the input changes *every* vertex is affected. Consequently, any update algorithm is a bounded algorithm, and Q is a bounded problem.

Showing that a problem Q is unbounded by reducing an unbounded problem P to Q involves the following obligations: (1) We must show how every instance of problem P (*i.e.*, the input) can be transformed into an instance of problem Q, and how the solution for this transformed problem instance can be translated back into a solution for the original problem instance. (2) We must show how any change δ_P to the original problem instance can be transformed into a corresponding change δ_Q in the target problem instance, and, similarly, how the change in the solution to the target problem instance can be transformed into the corresponding change in the solution to the original problem instance. (3) We must show that the time taken for the transformations referred to in (2) is *bounded* by some function of $\| \delta_P \|$. (4) We must show that $\| \delta_Q \|$ is also bounded by some function of $\| \delta_P \|$. (5) Finally, since we are dealing with the notion of unboundedness relative to the class of SA algorithms, we must show that the transformation algorithms referred to in (2) are SA algorithms.

Proposition 8.7. *Let $S = (S, \oplus, \otimes, \bar{0}, \bar{1})$ be an arbitrary closed semiring. The SS-S problem is unbounded for the class of SA algorithms.*

Proof. Given an instance of a single-source reachability problem (G, s), there is a linear-time reduction to an instance of SS-S given by $(G, s, \lambda e.\bar{1})$. In the target problem instance, the summary value at v, $C(s,v)$ is $\bar{1}$ if v is reachable from s, and $\bar{0}$ otherwise.

It is obvious that all the requirements laid down above for reduction among dynamic problems are met by the above reduction. Therefore, SS-S is an unbounded problem. $\square$

It follows from the above proposition that SSSP≥ 0 is an unbounded problem. In this particular problem the above reduction associates every edge in the graph with zero length. Consequently, the length of the shortest-path from the source vertex to a vertex u is 0 if the vertex is reachable, and ∞ otherwise.

However, as we saw in Section 3.1, the very similar problem SSSP>0 has a bounded locally persistent incremental algorithm. This illustrates that only certain input instances may be the reason why a problem is unbounded. For example, graphs with 0-length cycles cause SSSP≥ 0 to be unbounded. If the problematic input instances are unrealistic in a given application, it would be appropriate to consider a suitably restricted version of the problem that does not deal with these difficult instances.

8.4.2. Unbounded Dataflow Analysis Problems

In this section we show that all non-trivial meet-semilattice dataflow analysis problems are unbounded. Data-flow analysis problems are often cast in the following framework. (See Chapter 5.) The program gives rise to a *flow graph* G with a distinguished *entry* vertex s. Without loss of generality, s may be assumed to have no incoming edges. The problem requires the computation of some information $S(u)$ for each vertex u in the flow graph. The values $S(u)$ are elements of a meet semilattice L; a (monotonic) function $M(e):L \rightarrow L$ is associated with every edge e in the flow graph; and a constant $c \in L$ is associated with the vertex s. The desired solution $S(u)$ is the maximal fixed point of the following collection of equations:

$$S(s) = c$$
$$S(u) = \bigcap_{v \rightarrow u \in E(G)} M(v \rightarrow u)(S(v)), \quad \text{for } u \neq s.$$

Each semilattice L and constant $c \in L$, often the greatest or least element of the semilattice, determines a dataflow analysis problem, which we call the (L,c)-DFA problem. An input instance of the problem consists of a graph G and a mapping M from the edges of G to $L \rightarrow L$.

We now show that an arbitrary meet-semilattice dataflow analysis problem P is unbounded by reducing SS-REACHABILITY to P.

Proposition 8.8. *Let L be a meet-semilattice, and let $c \in L$. Then, the problem (L,c)-DFA is unbounded for the class of SA algorithms.*

Proof. Let f be a function from L to L such that $f(c) \neq \top$. Given an instance $((V,E),s)$ of the single-source reachability problem we can construct a corresponding instance $((V \cup \{t\}, E \cup \{(t \rightarrow s)\}),t,M)$ of problem P where,

$$M(e) = f \qquad \text{if } e = t \rightarrow s$$
$$M(e) = \lambda x.x \quad \text{if } e \neq t \rightarrow s.$$

The solution of this problem instance is given by: $S(t) = c$; if $u \neq t$, then $S(u)$ is $f(c)$ if u is reachable from s, and $\top$ otherwise. It follows from the unboundedness of SS-REACHABILITY that P is unbounded. $\square$

The interpretation of the above result is that any SA incremental algorithm for problem P is an unbounded algorithm. This does not by itself imply that the dataflow analysis problem P that arises in practice is an unbounded one for SA algorithms (in other words, if there is some flexibility in defining the class of valid input instances for problem P). The above reduction shows that some "difficult" input instances cannot be handled in time bounded by a function of $\|\delta\|$. However, these input instances may be unrealistic input instances in the context of the dataflow analysis problem under consideration. We now argue that, in fact, this is not the case.

The first possible restriction on input instances relates to the flow graph. Ordinarily, frameworks for batch dataflow analysis problems impose the assumption that all vertices in a flow graph be reachable from the graph's start vertex. Some dataflow analysis algorithms also assume that the flowgraph is a reducible one. With either of these restrictions on input instances, the above reduction of SS-REACHABILITY to problem P is no longer valid. However, we follow Marlowe [Mar89], who argued that these assumptions should be dropped for studies of incremental dataflow analysis (see Section 3.3.1 of [Mar89]).

The second possible restriction on input instances relates to the mapping M. Is it possible that realistic flow-graphs will never have a labeling corresponding to the "difficult" input instances shown to exist above? We argue below that this is not so.

The reduction above associated every edge with either the identity function or a function f such that $f(c) \neq \top$. The identity function is not an unrealistic label for an edge. (A **skip** statement, or more generally, any statement that modifies the state in a way that is irrelevant to the information being computed by the dataflow analysis problem P is usually associated with the identity function.) As for the function f, we now show that every non-trivial input instance must have an edge labeled by a function g such that $g(c) \neq \top$. Consider any input instance (G, s, M) such that $M(e)(c) = \top$ for every edge $e \in E(G)$. Since $M(e)$ must be monotonic, $M(e)(\top)$ must also equal $\top$. Then, the input instance (G, s, M) has the trivial solution given by:

$$S(s) = c$$
$$S(u) = \top \quad \text{for } u \neq s.$$

Hence, the edge-labeling M from the reduction used in the proof of Proposition 8.8 is, in fact, realistic.

In conclusion, note that the reduction used in the proof of Proposition 8.8 is independent of the class of incremental algorithms proposed (i.e., SA algorithms, LP algorithms, etc.). That is, the incremental version of every dataflow analysis problem is at least as hard as the dynamic single-source reachability problem. In other words, for a class of algorithms to have members that are bounded for any dataflow analysis

problem, there must be an algorithm of the class that solves the single-source reachability problem in a bounded fashion.

8.5. Some Remarks

There have been several papers on the dynamic reachability problem and the dynamic
transitive closure problem, but most of the previous work has been on incremental
algorithms for acyclic graphs, which suggests that the dynamic reachability problem
is difficult in the presence of cycles. The unboundedness result proved in this chapter
establishes, for the first time, that the dynamic reachability problem is hard, in some
sense. Similarly, the results in this chapter also show that incremental dataflow
analysis is hard, from the boundedness point of view, though a similar claim, from the
IRLB point of view (see Section 2.3.1.4), has been previously established by Berman
[Ber92].

The lower bounds have been established with respect to the sparsely-aliasing
pointer machine model. The lower bound also applies to *LP* algorithms (see
[Ram91, Ram96]). Neither the model of *LP* algorithms nor the model of *SA* algorithms allow the use of arrays (of 2 or higher dimensions) indexed by vertices. Some
examples of such arrays are the adjacency matrix and the transitive closure matrix. It
is possible to show that the unboundedness result holds for *LP* algorithms even if
multi-dimensional arrays indexed by vertices are allowed. Similarly, the proof in this
chapter can be adapted to show that the unboundedness result holds for *LP* algorithms
even when global auxiliary information, organized as in the pointer machine model, is
allowed subject to the following *separation* constraint: pointers from the local storage
(of *LP* algorithms) cannot point to the global store, and vice versa. What would be
desirable is a proof of the unboundedness result for arbitrary pointer machines or,
better still, for random access machines.

It is worth mentioning at this point that the restrictions of the models of computation we have used concern the organization of the auxiliary information that is
retained across input modifications. There is no restriction on the operations that can
be used *during* the update itself. Though these models do have some artificial constraints, they do suggest features an incremental algorithm *must* exploit to defeat the
lower bound proof.

> *I for one have to admit that I have not yet been able to devise a strict proof
> of this theorem ... As however the truth of it has been established in so many
> cases, there can be no doubt that it holds good for any solid. Thus, the proposi-
> tion seems to be satisfactorily demonstrated.*
>   Euler, concerning his conjecture that V−E+F = 2 for polyhedrons [as cited in]
>   —I. Lakatos, *Proofs and Refutations: The Logic of Mathematical Discovery*

Chapter 9
Incremental Algorithms for Reducible Flowgraphs

9.1. Introduction

In this chapter we present incremental algorithms for the single-source reachability problem and the dominator tree problem in reducible flowgraphs. Both these algorithms are unbounded algorithms.[1] The reachability algorithm, with a running time of $O(\|\delta\| \log n)$, is an interesting example of an algorithm with a "hybrid complexity measure". The reachability algorithm can be adapted to work for arbitrary graphs, but the time complexity does not carry over to the case of irreducible graphs.

9.2. Reachability, Domination, and Reducible Flowgraphs

The concepts of reachability, domination, and reducible flowgraphs are closely related to each other. We first review the concepts of domination and reducibility [Aho86].

A *flowgraph* is a directed graph with a distinguished source vertex. We initially consider flowgraphs in which every vertex is reachable from the source vertex. A vertex u is said to *dominate* a vertex v in a flowgraph with source s iff every path from s to v passes through u. Note that every path from s to v contains u iff every acyclic path from s to v contains u; consequently, to check that v dominates u it is sufficient to examine the acyclic paths from s to u. Domination is a reflexive and transitive relation—a vertex u dominates itself and if u dominates v and v dominates w then u dominates w. If u dominates v and $u \neq v$ then u is said to be a *proper dominator* of v.

Domination is a special kind of relation. If u and v are two dominators of a vertex w then one of u and v must dominate the other. Consequently, every vertex u has an *immediate dominator* v such that any proper dominator of u is a dominator of v. We denote the immediate dominator of u by $idom(u)$. The *dominator tree* of a flowgraph is a tree consisting of all the vertices reachable from the source s constructed as follows: The source vertex s is made the root of the dominator tree, and every reachable vertex u, other than s, is made a child of its immediate dominator. The dominator tree is a concise representation of the domination relation—a vertex u dominates another vertex v iff u is an ancestor of v in the dominator tree.

---

[1]We saw in the previous chapter that the reachability problem is unbounded with respect to several models of computation. Reps (personal communication) has shown that the proof of unboundedness of reachability with respect to the class of LP algorithms can be adapted for the dynamic dominator tree problem too. However, these lower bound proofs involve irreducible flowgraphs, and do not apply to the problems considered here. Nevertheless, the algorithms presented in this chapter show what one may be able to do in the case of unbounded problems.

An edge $x \rightarrow y$ in a flowgraph is said to be a *back edge* if vertex y dominates vertex x and it is said to be a *forward edge* otherwise. A flowgraph is said to be a *reducible flowgraph* if the set of all forward edges induces an acyclic graph. Otherwise, it is said to be *irreducible*.

The normal definition of a reducible flowgraph assumes that all vertices are reachable from the source vertex. We relax this restriction and consider a flowgraph to be a reducible flowgraph if the set of vertices reachable from the source is a reducible flowgraph according to the above definition. This relaxation is useful in the context of incremental computation—in changing one reducible flowgraph into another reducible flowgraph by inserting and deleting edges it may be necessary to temporarily introduce unreachable vertices. However, the concept of domination still applies only to reachable vertices. In particular, the algorithm we outline in this chapter will maintain the dominator tree of the subgraph induced by the reachable vertices.[2] Also, the classification of an edge as a back or forward edge is meaningful only if the endpoints of the edge are reachable. By the "status" of an edge we mean its status as a back or forward edge.

Recall that an edge $u \rightarrow v$ is a back edge iff v dominates u. Thus, any path from the source vertex s that contains a back edge $u \rightarrow v$ must be a cyclic path since it contains at least two occurrences of the vertex v. This implies that the set of reachable vertices does not change if a back edge $u \rightarrow v$ is removed from the graph, since a vertex is reachable iff there exists an acyclic path from the source vertex to that vertex. Similarly, the domination relation of a flowgraph does not change if a back edge is removed from the graph, since domination can be defined in terms of acyclic paths. Hence, the problems of maintaining the reachability information and the dominator tree of a reducible flowgraph are closely related to the problems of maintaining the reachability information and the dominator tree of a dag (the dag of forward edges), provided we have a way of identifying the forward edges and back edges of the graph. We now show that the status of an edge as a forward edge or back edge does not change as edges are inserted into and deleted from the graph as long as the graph remains reducible throughout this sequence of modifications. (We are talking only of edges whose endpoints remain reachable in the flowgraph.)

Proposition 9.1. Let G_1 and G_2 be two reducible flowgraphs such that $E(G_2) = E(G_1) \cup \{u \rightarrow v\}$. The status of an edge that is reachable in both graphs is the same.

---

[2]Carroll [Car88a] extends the definition of dominator tree to that of a *dominator forest*, in the presence of unreachable vertices: he considers a decomposition of the whole graph into a collection of flowgraphs, each with its own source vertex, such that every vertex is reachable from the source of the flowgraph it belongs to; the collection of the dominator trees of these flowgraphs constitute a dominator forest. If the graph has a unique minimal decomposition, then this approach is meaningful. However, a graph need not have a unique minimal decomposition, in general, and the dominator forest is not uniquely defined. Since the advantages of maintaining such a dominator forest are unclear, we restrict our attention to the problem of maintaining the dominator tree of the reachable vertices.

Proof. It follows easily from the definition of domination that the insertion of an edge can only shrink the domination relation—hence, if x dominates y in graph G_2 then x must dominate y in graph G_1 too. Consequently, a back edge in graph G_2 must be a back edge in graph G_1 too (assuming that the edge under consideration is not $u \rightarrow v$). Equivalently, every forward edge in G_1 is a forward edge in G_2.

Going the other way, a back edge in G_1 must be a back edge in G_2 also, for the following reason: if $x \rightarrow y$ is a back edge in G_1, then there is some path consisting only of forward edges from y to x in G_1; hence, there exists a path of forward edges from y to x in G_2 also; if $x \rightarrow y$ were a forward edge in G_2, then the set of forward edges in G_2 would induce a cycle, contradicting the assumption that G_2 is reducible. $\square$

Let us now consider the status of a newly inserted edge $u \rightarrow v$. Since domination can be defined in terms of acyclic paths, insertion of an edge $u \rightarrow v$ does not change the set of dominators of vertex u. Consequently, $u \rightarrow v$ is a back edge iff v dominates u in the new graph iff v dominates u in the old graph. This is useful since we can determine the status of a newly inserted edge from the domination information about the original graph.

9.3. The Dynamic Single-Source Reachability Problem

The problem addressed here is the following special case of the dynamic single-source reachability problem: the set of all vertices reachable from a distinguished vertex (the "source") of a given graph is to be maintained, as the given graph undergoes modifications such as the insertion or deletion of an edge, subject to the constraint that at all times the set of vertices reachable from the source induces a reducible flowgraph. We will later discuss how this assumption can be relaxed and also how one can verify that an input modification does not violate this requirement.

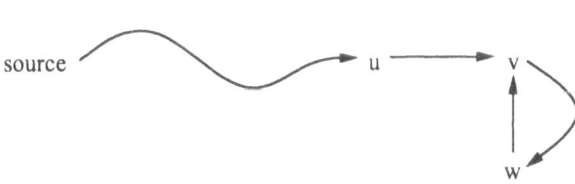

Figure 9.1. An example to show how the presence of cycles complicates determining if the deletion of an edge $u \rightarrow v$ leaves vertex v unreachable. It is not sufficient to check if v has some predecessor marked reachable. We need to determine if v has some predecessor that is marked reachable *and is not dominated by* v. All vertices in the above example are marked reachable to begin with.

Most previous work on the dynamic reachability problem or the dynamic transitive closure problem has been on restricted versions of these problems—either only edge insertions where allowed, or the input graph was restricted to be a dag throughout the sequence of modifications. The most difficult aspect of the dynamic reachability problem is, in fact, handling edge deletions in the presence of cycles in the graph. The only previous algorithm that handles both edge deletions and cyclic graphs is due to La Poutre and van Leeuwen [Pou88]. This algorithm maintains a strongly connected component decomposition of the graph in order to maintain the transitive closure of the graph. This algorithm has been analyzed separately for a sequence of edge insertions and a sequence of edge deletions, making it difficult to make direct comparisons, but the algorithm appears to be more expensive than the algorithm outlined in this section in the case of reducible flowgraphs. The other dynamic reachability algorithms work only for acyclic graphs and are based on the ideas reproduced below.

Updating the reachability information is reasonably simple when an edge $u \rightarrow v$ is inserted. Nothing needs to be done if u is unreachable or v is already reachable. Otherwise, v should be marked as being reachable, and the "reachable" status should be propagated further by performing a graph traversal from v, taking care to avoid visiting successors of vertices that were already reachable. This, in fact, takes time $O(\|\delta\|)$.

Let us now consider the difficulty in updating the reachability information after the deletion of an edge $u \rightarrow v$. The question to be answered is if the deletion of the edge $u \rightarrow v$ makes v unreachable. If this can be answered, then the reachability information for the whole graph can be updated by essentially repeatedly asking questions of this form, as one propagates the "unreachable" status from v, stopping the propagation whenever one reaches a vertex that does not become unreachable.

Let us first consider the case of acyclic graphs, where this question has a simple answer. The deletion of an edge $u \rightarrow v$ from a dag leaves v unreachable iff v has no predecessors (other than u) that are marked reachable. If we determine that v has become unreachable then we would mark it so and then examine its successors to see if any of them has become unreachable. Change propagation would proceed in the obvious way. This leads to a linearly bounded incremental algorithm if carefully implemented: if we maintain a count, at each vertex, of the number of reachable predecessors it has, then visiting a vertex is a constant time operation, and the algorithm runs in $O(\|\delta\|)$ time. Of course, this would imply that both the edge-insertion and the edge-deletion algorithm must update the counts appropriately, but that does not add to the algorithm's asymptotic complexity. Observe that this is essentially a simplified version of Phase 1 of algorithm DeleteEdge$_{SSSP>0}$ (see Figure 4.2), where we were interested in the sub-problem of identifying the vertices that could not reach the sink after the deletion of an edge in the shortest-paths subgraph (which is a dag).

The above scheme for checking if a vertex has become unreachable fails once we allow cycles in the graph. If, after the deletion of an edge $u \rightarrow v$, the vertex v has no predecessors marked reachable then v, in fact, is unreachable. *But the converse is not true: even if v has some predecessor w marked reachable we cannot be sure that v*

is reachable. In the graph shown in Figure 9.1, for example, v does have a predecessor w marked reachable, but all paths to w go through the deleted edge $u \rightarrow v$. What we need to check, in the presence of cycles, is if v has some predecessor w, marked reachable, *that is not dominated by* v. We may restate this as: the deletion of edge $u \rightarrow v$ leaves v unreachable iff v has no incoming forward edge $w \rightarrow v$ such that w is marked reachable. Consequently, this check can be done easily if we can determine which of the edges in the graph are forward edges and which are back edges.

 We now present a simple test for checking if an edge in a reducible flowgraph is a forward edge or a back edge.

Proposition 9.2. Let T be any directed spanning tree of a reducible flowgraph rooted at the source vertex. (The edges are directed from the parent to the child.) An edge $x \rightarrow y$ is a back edge in the given graph iff vertex x is a descendant of vertex y in T.

Proof. Assume that $x \rightarrow y$ is a back edge. Then, y dominates x, from the definition of a back edge. Hence, any path from the source to x must contain y. Consequently, x must be a descendant of y in any spanning tree of the graph.

 Assume that x is a descendant of y in the spanning tree. Now, any spanning tree edge $u \rightarrow v$ is a forward edge, since v does not dominate u. Hence, there exists a path of forward edges from y to x. If $x \rightarrow y$ were a forward edge, then the forward edges would induce a cycle in the graph. Hence $x \rightarrow y$ must be a back edge. $\square$

 We now present our algorithm for maintaining reachability information in a reducible flowgraph, based on the ideas outlined above. The algorithm works by maintaining a spanning tree of the set of all vertices reachable from the source using the link-cut tree data structure due to Sleator and Tarjan [Sle83]. This dynamic data structure is a representation of a dynamic forest (collection of trees) that allows a number of operations to be performed. The only operations that are of interest to us are: a *maketree* operation that creates a new tree consisting of a single newly inserted vertex; a *link* operation that adds an edge from some vertex u in a tree T_1 to the root of another tree T_2, making T_2 a subtree of T_1; a *cut* operation that removes an edge from a tree, breaking it into two trees; an operation that checks if a vertex u is the descendent of another vertex v. (The link-cut tree data structure supports an operation that returns the least common ancestor of two vertices. This can be directly used to check for the ancestor-descendant relationship between two vertices.) Each of these operations runs in $O(\log n)$ time, where n is the number of vertices in the forest.

 Given a spanning tree of the reachable vertices in a flowgraph, let us say that u is a *support vertex* for v if u is a reachable predecessor of v and u is not a descendant of v in the spanning tree. It follows from the above discussion that u is a support vertex of v in a reducible flowgraph iff u is a reachable vertex and $u \rightarrow v$ is a forward edge. Thus, the set of support vertices of a vertex v is the same for every spanning tree in the case of reducible flowgraphs. This is not true for irreducible graphs.

 We previously saw that maintaining a count of the number of reachable predecessors of a vertex was useful in updating the reachability information in the case of dags. Similarly, we will find it useful, in the case of reducible flowgraphs, to maintain the set of support vertices of every vertex—this is not necessary, but it improves

procedure InsertEdge$_{Reachability}$($G, T, u \rightarrow v$)
declare
      G: a directed graph
      T: a link-cut tree
      $u \rightarrow v$: an edge to be inserted into G
      WorkSet: a set of edges
preconditions
      $v \rightarrow w \notin E(G)$
      T is a spanning tree of the reachable vertices in G
      $\forall v \in V(G)$, *support*(v) is the set of support vertices for v
      $\forall v \in V(G)$, *reachable*(v) is true iff v is reachable
begin
[1]   Insert edge $u \rightarrow v$ into $E(G)$
[2]   **if** *reachable*(u) **then**
[3]   WorkSet := $\{u \rightarrow v\}$
[4]   **while** WorkSet $\neq \emptyset$ **do**
[5]   Select and remove an edge $x \rightarrow y$ from WorkSet
[6]   **if** not *reachable*(y) **then**
[7]   *support*(y) := $\{x\}$
[8]   *reachable*(y) := *true*
[9]   *link*(x,y) in T
[10]   **for** every vertex $z \in$ *Succ*(y) **do**
[11]   Insert $y \rightarrow z$ into WorkSet
[12]   **od**
[13]   **else if** x is not a descendant of y in T **then**
[14]   /* $x \rightarrow y$ is a forward edge */
[15]   Add x to *support*(y)
[16]   **fi**
[17]   **od**
[18]   **fi**
end
postconditions
      $v \rightarrow w \in E(G)$
      T is a spanning tree of the reachable vertices in G
      $\forall v \in V(G)$, *support*(v) is the set of support vertices for v
      $\forall v \in V(G)$, *reachable*(v) is true iff v is reachable

Figure 9.2. An algorithm to update the reachability information after the insertion of an edge $v \rightarrow w$ into graph G.

the time complexity of the updating algorithms from $O(\|\delta\|_2 \log n)$ to $O(\|\delta\| \log n)$. Note that a vertex is reachable iff its set of suppport vertices is non-empty. Hence, we don't need to maintain a separate "reachability" status flag for every vertex, but we do so for readability.

9.3.1. Insertion of an Edge

We now consider how the algorithm processes the insertion of an edge $u \rightarrow v$ into the graph. See Figure 9.2. If the vertex u is currently unreachable from the source, then nothing needs to be done. Otherwise, a work-list is created, initially consisting of the edge $u \rightarrow v$ alone. The algorithm repeatedly extracts an edge $x \rightarrow y$ from the work-list and processes it as follows. If y was previously unreachable then y is marked as being reachable, *support*(y) is appropriately initialized, the edge $x \rightarrow y$ is added to the spanning tree, and all edges $y \rightarrow z$ going out of vertex y are added to the work-list. If the vertex y is already reachable from the source, then not much needs to be done—if x is not a descendant of y then x is a new support vertex for y, so we add it to *support*(y).

Note that for every edge $x \rightarrow y$ examined by the algorithm, x must be in CHANGED. Adding an edge $x \rightarrow y$ to the spanning tree takes $O(\log n)$ time, since the spanning tree is maintained as a link-cut tree. Similarly, the test in line [13] also takes $O(\log n)$ time. Hence, the insertion of an edge is processed in time $O(\|\delta\| \log n)$ time.

9.3.2. Deletion of an Edge

We now consider how the deletion of an edge $u \rightarrow v$ is processed. See Figure 9.3. A work-list is created, initially consisting of the edge $u \rightarrow v$ alone. The algorithm repeatedly extracts an edge $x \rightarrow y$ from the work-list and processes it as follows. If x is a support vertex for y then it is removed from *support*(y). If $x \rightarrow y$ is a spanning tree edge then it is removed from the spanning tree using a cut operation. This gives us the subtree T of the original spanning tree rooted at vertex y. We check y to see if there exists any incoming forward edge $z \rightarrow y$ such that z is marked reachable— recall that such an edge exists iff *support*(y) is non-empty. If such a vertex z exists, then we link z and y, *i.e.*, y is made a child of z in the spanning tree, and the tree T itself becomes a subtree of the tree rooted at z. If such a vertex z does not exist, then y is marked as being unreachable from the source, and all edges $y \rightarrow z$ going out of vertex y are added to the work-list.

Note that for every edge $x \rightarrow y$ in the work list, the vertex x is in CHANGED. Consequently, the algorithm runs in time $O(\|\delta\| \log n)$.

9.3.3. Handling Irreducible Flowgraphs

Let us now consider irreducible graphs. Note that only the insertion of an edge can introduce irreducibility—the deletion of an edge from a reducible flowgraph leaves a reducible flowgraph. If it is necessary to check if a flowgraph remains reducible during a sequence of modifications, we can do so by verifying that the subgraph induced by the forward edges remains acyclic. This can be done, for instance, by maintaining a priority ordering of the dag of forward edges, as discussed in the Section 9.4. We now briefly show how the above incremental algorithms can be generalized to maintain reachability information even for irreducible flowgraphs. The generalized algorithms, however, may not perform the updating in time $O(\|\delta\| \log n)$ when the input graph is irreducible.

procedure DeleteEdge$_{Reachability}$(G, T, $u \rightarrow v$)
declare

        G: a directed graph

        T: a link-cut tree

        $u \rightarrow v$: an edge to be deleted from G

        WorkSet: a set of edges

preconditions

        $v \rightarrow w \in E(G)$

        G is a reducible flowgraph

        T is a spanning tree of the reachable vertices in G

        $\forall v \in V(G)$, $support(v)$ is the set of support vertices for v

        $\forall v \in V(G)$, $reachable(v)$ is true iff v is reachable

begin

[1]   Remove edge $u \rightarrow v$ from $E(G)$

[2]   **if** $reachable(u)$ **then**

[3]   WorkSet := { $u \rightarrow v$ }

[4]   **while** WorkSet $\neq \varnothing$ **do**

[5]   Select and remove an edge $x \rightarrow y$ from WorkSet

[6]   **if** x is not a descendant of y in T **then**

[7]   Remove x from $support(y)$

[8]   **fi**

[9]   **if** $x \rightarrow y$ is a spanning tree edge in T **then**

[10] $cut(x,y)$

[11]   **if** $support(y) = \varnothing$ **then**

[12] $reachable(y) := false$

[13]   **for** every vertex $z \in Succ(y)$ **do**

[14]   Insert $y \rightarrow z$ into WorkSet

[15]   **od**

[16]   **else**

[17]   Choose some z from $support(y)$

[18] $Link(z,y)$ in T

[19]   **fi**

[20]   **fi**

[21]   **od**

[22]   **fi**

end

postconditions

        $v \rightarrow w \notin E(G)$

        T is a spanning tree of the reachable vertices in G

        $\forall v \in V(G)$, $support(v)$ is the set of support vertices for v

        $\forall v \in V(G)$, $reachable(v)$ is true iff v is reachable

Figure 9.3. An algorithm to update the reachability information after the deletion of an edge $v \rightarrow w$ from graph G.

Let us forget domination, forward edges, and reducibility, and summarize the previous algorithm as follows. It maintains a spanning tree T of the reachable vertices. It also maintains the set of support vertices of every vertex, where u is a support vertex for v if u is a reachable predecessor of v and u is not a descendant of v in the spanning tree.

It can be verified that the algorithm for edge-insertion, in fact, works correctly even if the input graph is irreducible, and takes only time $O(\|\delta\| \log n)$. But there are two problems in handling an edge-deletion. The first problem is that the algorithm is based on the assumption that a vertex is unreachable if its support set becomes empty and this assumption is not justified in the case of irreducible graphs. It is true, however, that a vertex is reachable as long as its support set is non-empty. Consequently, every vertex labelled reachable at the end of the updating must be reachable, but some of the vertices labelled unreachable may be reachable.

We fix this problem by modifying the edge-deletion algorithm as follows. Consider line [12] where vertex y is marked as being unreachable. Let S denote the set of predecessors of y that are currently marked reachable. Since we know at this point that y has no support vertex, every vertex in S must be a descendant of y in the current spanning tree. We mark y as being unreachable *assuming that every vertex in S is, in fact, unreachable*—this assumption is true in the case of reducible flowgraphs, but may not be true for an irreducible flowgraph. (See Figure 9.4 for an example.) So, we maintain a set *AssumedUnreachable*, the set of all vertices that have been assumed to be unreachable. At line [12] we add all vertices in S to this set. At the

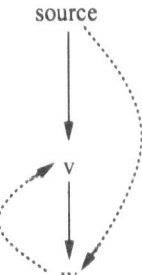

source

Figure 9.4. An irreducible flowgraph. The solid edges represent the edges of a spanning tree. Note that v remains reachable even after the deletion of the edge from *source* to v. The edge-deletion algorithm for reducible flowgraphs will, however, mark v as having become unreachable. This happens because the only edge coming into v, after the deletion of *source* $\rightarrow v$, comes from a descendant of v in the (original) spanning tree. Thus, v is marked unreachable under the assumption that w too has become unreachable. However, w is not marked unreachable when the updating ends (because it has an incoming edge *source* $\rightarrow w$). This provides a way of checking if the algorithm updated the reachability information erroneously because of irreducibility.

end we compute *FalselyAssumedUnreachable*, the set of all vertices in *AssumedUn-reachable that were not marked as having become unreachable during the updating.* If this set is not empty, then we made a mistake. So, we perform a traversal starting from the vertices in *FalselyAssumedUnreachable* and propagate the "reachable" status (to all wrongly marked vertices). This traversal undoes any mistakes made earlier. In the case of reducible flowgraphs, *FalselyAssumedUnreachable* will be empty, and there is no asymptotic increase in the complexity of the updating. If the graph is irreducible, however, the algorithm may have examined more than $\| \delta \|$ vertices. In the worst case, the algorithm might perform $\Theta(m \log n)$ work.

A second problem with handling edge-deletion in the case of irreducible flowgraphs is that maintaining the support sets is no longer easy—when we change the spanning tree by moving subtrees in lines [17-18] the support sets can change drastically if the graph is irreducible. So, we avoid maintaining support sets, and modify lines [11] and [17] to scan the predecessors of y to determine if it has a support vertex. This has to be done carefully, since the vertex y might be "examined" in this fashion several times. In particular, we should ensure that in future visits of vertex y, any predecessor of y that has already been checked (to see if it is a support vertex for y) is not checked again. If we take care of this, the deletion of an edge can be processed by the algorithm in time $O(\| \delta \|_2 \log n)$ in the case of reducible flowgraphs.

9.4. The Dynamic Dominator Tree Problem

In this section we present an incremental algorithm for maintaining the dominator tree of the subgraph of reachable vertices of a flowgraph under the assumption that this subgraph remains reducible as the flowgraph undergoes modifications. We also outline a method for verifying that the graph, in fact, remains reducible during a sequence of modifications. (The material in this section has previously appeared in [Ram94].)

The dominator tree plays an important role in several algorithms for program analysis and program optimization, and the need for updating the dominator tree of a flowgraph arises in various contexts. For instance, Carroll and Ryder [Car88] present an incremental dataflow analysis algorithm that makes use of dominator trees—it is necessary as a part of this algorithm to update the dominator tree of the flowgraph. The need to update the dominator tree can arise even in the context of batch compilation. For instance, dominator trees are used in the construction of the static single assignment (SSA) representation of programs [Cyt89, Cyt91]. As an optimizing compiler repeatedly applies optimizing transformations, it may be necessary to update the SSA representation of the program after each transformation.

The only previous algorithm for the problem of maintaining the dominator tree of a (reducible) flowgraph is due to Carroll and Ryder [Car88]. The algorithm given in this section has a better worst-case complexity than the Carroll-Ryder algorithm. There are also good reasons to believe that our algorithm will be more efficient in practice also. A comparison of the two algorithms is given later in Section 9.4.3.

We saw in Section 9.2 that the dominator tree of a reducible flowgraph is the same as that of the dag of forward edges of that flowgraph. We will first present a simple batch algorithm for constructing the dominator tree of a dag. Linear time algorithms that construct the dominator tree of a dag are known [Har85, Och83], but the reason for the following presentation is that it suggests a possible way of incrementally maintaining the dominator tree of a dag. We will then use this idea for maintaining the dominator tree of a reducible flowgraph.

Consider a dag with a source vertex. Consider a vertex u in the dag with predecessors $v_1, \ldots, v_k$. A vertex w will properly dominate u iff it dominates all the vertices v_1 through v_k. In other words, $Dom(u) = \{u\} \cup \bigcap_{i=1}^{k} Dom(v_i)$, where $Dom(x)$ denotes the set of dominators of vertex x. Thus, if we have identified the set of dominators of v_1 through v_k, then the set of proper dominators of u can be obtained by intersecting these sets. In particular, the immediate dominator of u has to be the least common ancestor of v_1 through v_k in the dominator tree.

The previous paragraph suggests the following scheme for constructing the dominator tree of a dag in an incremental fashion. The vertices in the dag are visited in topological order and added to the dominator tree one by one. Initially the dominator tree consists only of the source vertex s. When a vertex u is visited, the least common ancestor w of all the predecessors v_1 through v_k of u in the partially constructed tree is identified. Now, u is made a child of w. Thus, the construction of the dominator tree of a dag can be considered to be a "pseudo" circuit value annotation problem, where every vertex u other than the source vertex is associated with the equation

$$parent(u) = lca(v_1, \ldots, v_k),$$

where we use lca as an abbreviation for "least common ancestor". The value computed for every vertex (other than the source) is its parent in the dominator tree.

Note that the above problem is not a true circuit value annotation problem because the right-hand side of the above equation is not a "pure function" of the values associated with the predecessors of vertex u. Hence, the output value of a vertex (that is, its immediate dominator) can change even though the output value of none of its predecessors have changed. Consider, for example, the dag G shown in Figure 9.5. This dag is modified into dag G' by the insertion of an edge $b \rightarrow c$. The vertex c is affected in the sense that its immediate dominator changes. The only successor of c, namely d, is not affected—its immediate dominator is still c. But, vertex e, is affected, even though none of its predecessors are affected.

The above problem prevents us from using change-propagation techniques in updating the dominator tree after the insertion or deletion of an edge. However, if we have a conservative approximation to the set of all affected vertices, then we can use ideas from the circuit value annotation problem to update the dominator tree. In particular, we can visit all vertices that might possibly be affected in topological order, and determine their immediate dominators using the above equation. We can use priority-ordering to visit vertices in a topological sort order, and use the incremental algorithm of Alpern et al. [Alp90] for updating the priority ordering when the graph

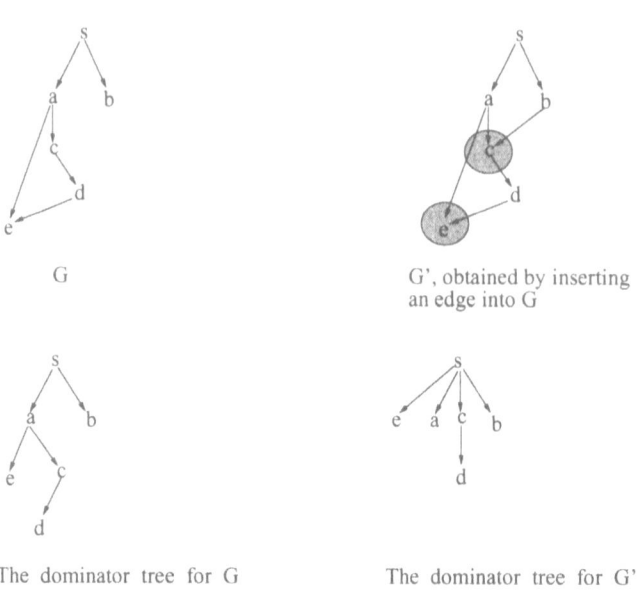

Figure 9.5. An example to show how the problem of constructing the dominator tree of a dag differs from a circuit value annotation problem. An edge $b \rightarrow c$ is inserted into the dag G. The affected vertices in the resulting dag G', indicated by the shaded region, do not form a connected region.

undergoes changes. We can represent the dominator tree using the link-cut tree data structure, since this data structure lets us compute least common ancestors of vertices in dynamic trees efficiently.

We now return to the problem of maintaining the dominator tree of a reducible flowgraph. Let G denote the reducible flowgraph for which the dominator tree has to be maintained. The algorithm will maintain the following data structures and information: (a) The reachability status of every vertex will be maintained. (b) Let $R(G)$ denote the subgraph induced by the reachable vertices. The status of every edge in $R(G)$—whether it is a forward edge of back edge—will be maintained. (c) Let $F(G)$ denote the acyclic subgraph of $R(G)$ induced by the forward edges. Both $F(G)$ and a correct prioritization of $F(G)$ will be maintained. (d) The dominator tree $DT(G)$ of $F(G)$, which is the same as the dominator tree of $R(G)$ and G, will be maintained as a link-cut tree.

9.4.1. Insertion of an Edge

We now consider the problem of updating all the above information when an edge $u \rightarrow w$ is inserted into the graph. We will assume that the vertex u was originally reachable from the source vertex, since nothing needs to be done otherwise. We will first consider the simpler case where the vertex w was already reachable, which means that there is no change in the reachability status of vertices.

A Special Case: No Change in Reachability

In this case the set of vertices in $R(G)$ remains the same, while $u \rightarrow w$ is added to the set of edges. We know from Proposition 9.1 that the insertion of $u \rightarrow w$ does not change the status of any other edge in $R(G)$. Consequently, maintaining edge statuses requires only that we determine the status of the newly inserted edge. If vertex w dominates vertex u in the original graph, then the edge inserted is a back edge. Otherwise, it is a forward edge. We can check if v dominates u in the original graph in $O(\log n)$ time, since a representation of the dominator tree as a link-cut tree enables us to check for the ancestor-descendant relation between two vertices in $O(\log n)$ time.

   If the edge $u \rightarrow w$ is a back edge, then neither $F(G)$ nor $DT(G)$ changes, and nothing more needs to be done. If the edge is a forward edge, then we insert it into $F(G)$, and update the prioritization of $F(G)$ using the algorithm presented in [Alp90]. Updating the priorities takes $O(\gamma \log \gamma)$ time, where γ is a measure of the size of the change in the priorities.

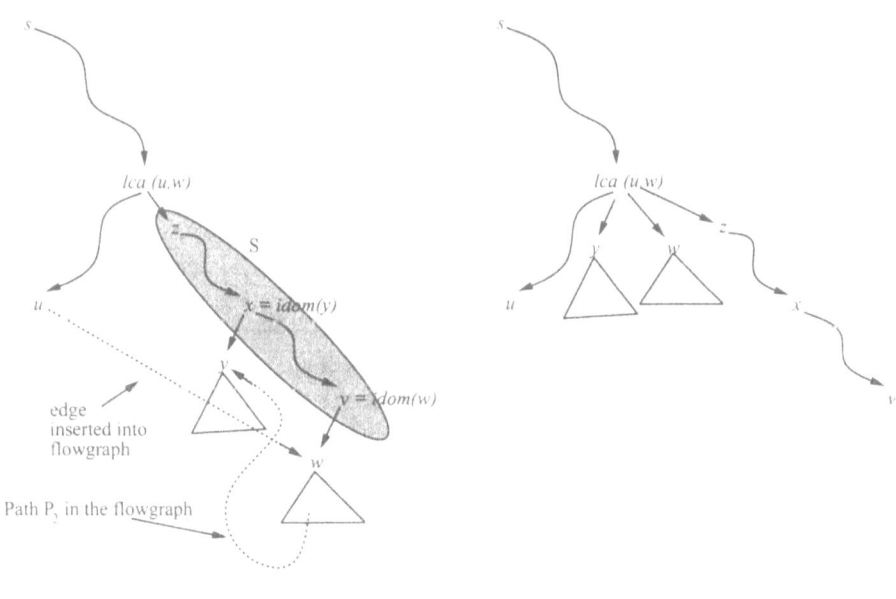

Figure 9.6. The change in the dominator tree of a flowgraph when an edge $u \rightarrow w$ is inserted. Solid wavy lines in the figure indicate paths in the dominator tree, while solid straight lines indicate edges in the dominator tree. Dashed lines indicate edges and paths in the flowgraph. For any affected vertex y, $idom(y)$ must lie in the shaded region of the original dominator tree. Further, the new immediate dominator of any affected vertex is $lca(u,w)$.

We now consider the problem of updating $DT(G)$ itself. Let us now consider how the dominator tree can change following the insertion of a forward edge into the flowgraph.

Proposition 9.3. (See Figure 9.6.) Consider the insertion of an edge $u \longrightarrow w$ into a flowgraph where both u and w are already reachable. If a vertex y is affected by the insertion of the edge, then $idom(y)$ must properly dominate w and $idom(y)$ must be properly dominated by $lca(u,w)$. Further, the new immediate dominator of every affected vertex must be $lca(u,w)$.

Proof. We have assumed that vertex w was already reachable in the flowgraph. Hence, w must occur in the dominator tree of the original flowgraph. Let v denote $idom(w)$. Consider the least common ancestor of u and w. Consider an affected vertex y, that is, a vertex y whose immediate dominator changes. Let x denote $idom(y)$ in the original graph. We noted earlier that the insertion of an edge can only shrink the domination relation. Hence, y's new immediate dominator must have been a dominator of y in the original graph too—that is, y's new immediate dominator must be an ancestor of y in the original dominator tree.

What can we infer from the fact that x no longer dominates y? The insertion of the edge $u \longrightarrow w$ must have created a path from s to y that avoids x. It follows that the original graph must contain a path P_1 from s to u and a path P_2 from w to y both of which avoid vertex x. This implies that x cannot have been a dominator of u in the original graph, since otherwise the required path P_1 could not have existed. This also implies that x must have been a proper dominator of w in the original graph— otherwise, there must exist a path P_3 in the original graph from s to w that avoids x. Concatenating paths P_3 and P_2 yields a path from s to y in the original graph that avoids x, contradicting the assumption that x dominates y in the original graph.

Hence, x must be a proper ancestor of w, but cannot be an ancestor of u. In other words, x must be a proper ancestor of w and a proper descendant of $lca(u,w)$— that is, it must lie in the shaded region S shown in Figure 9.6. This establishes the first claim in the proposition.

Now, consider the second claim. Assume that y is an affected vertex. Note that $lca(u,w)$ dominates y even in the new graph, since there exists no path from s to u that avoids $lca(u,w)$. On the other hand, no ancestor t of y in the original domina- tor tree that is a proper descendant of $lca(u,w)$ can dominate y in the new graph— there exists a path P_2 from w to y that avoids t (since y is assumed to be an affected vertex), and there exists a path P_1 from s to u that avoids t. Consequently, the new immediate dominator of an affected vertex y must be $lca(u,w)$. □

For any two vertices p and q define *PossiblyAffected* (p,q) to be the set $\{\, r \mid idom(r)$ is a proper ancestor of q and a proper descendant of $lca(p,q)\,\}$. The above proposition shows that *PossibleAffected* (u,w) is a conservative approximation to the set of affected vertices when an edge $u \longrightarrow w$ is inserted, provided both u and w are reachable in the original graph. We can "re-evaluate" all these vertices in increasing order of priority to determine the set of affected vertices and to update the dominator

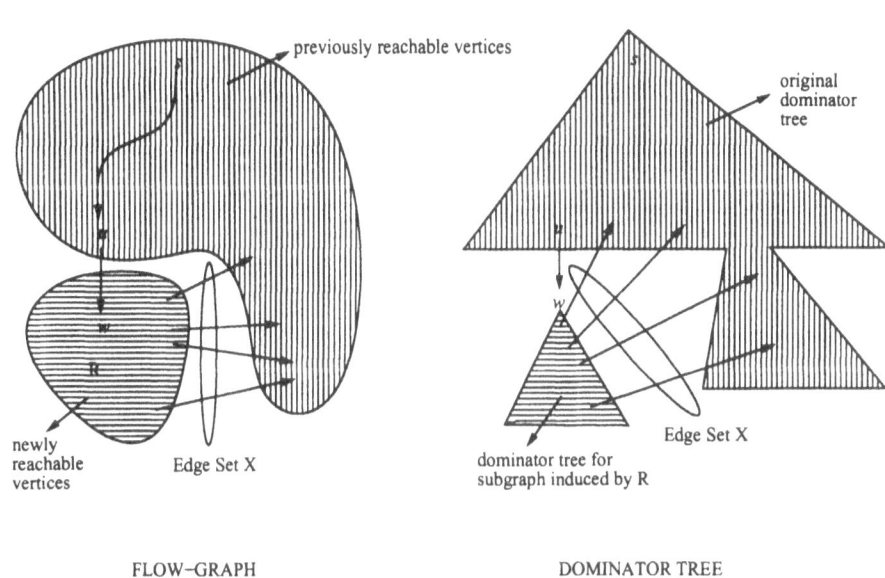

previously reachable vertices

original dominator tree

newly reachable vertices

Edge Set X

Edge Set X

dominator tree for subgraph induced by R

FLOW–GRAPH   DOMINATOR TREE

Figure 9.7. The change in the dominator tree of a flowgraph when an edge $u \rightarrow w$ is inserted such that w becomes reachable. The flowgraph is shown on the left side. The tree on the right side shows how the new dominator tree would look like *if there was no edge from a newly reachable vertex to a previously reachable vertex—that is, if set X was empty.* If X is not empty, then the new dominator tree can be obtained from the dominator tree on the right by "processing the insertion of edges in F".

tree. The "re-evaluation" of vertices in *PossibleAffected* (u,w) is necessary only to determine the vertices that are actually affected, since, as the above proposition shows, the new immediate dominator of every affected vertex y is $lca(u,w)$. This step takes time $O(\parallel PossibleAffected(u,w) \parallel \log n)$.

The General Case

Consider the insertion of an edge $u \rightarrow w$ where u is reachable but w may or may not be reachable. This can be processed very easily using a worklist algorithm that repeatedly invokes the special case updating algorithm, as follows. Create a worklist that initially consists of just $u \rightarrow w$. Every edge $x \rightarrow y$ in the worklist will be processed as follows: If y is already marked reachable, we process edge $x \rightarrow y$ using the special case algorithm previously outlined. (Except for the particular case where the edge $x \rightarrow y$ is the newly inserted edge $u \rightarrow w$, the edge $x \rightarrow y$ would have already been in the graph, but would not have been previously processed since vertex x was previously unreachable. Hence it is being processed now.) If y is marked unreachable then we do the following: y is marked reachable; the edge $x \rightarrow y$ is marked as a forward edge; y is assigned a priority greater than x; y is made a child of vertex x in the dominator tree; and, finally, all the edges going out of y are added to the worklist.

However, it is possible to do the update somewhat more efficiently in the general case. Observe that the edges processed in the above worklist algorithm are of two types: edges of the form $x \longrightarrow y$ where y was already reachable, and edges of the form $x \longrightarrow y$ where y was not previously reachable. The following algorithm first identifies all the edges and vertices that need to be processed, and partitions them into these two classes, and processes them separately.

An outline of the algorithm is presented in Figure 9.8. The major steps involved in the algorithm are: (1) Determining the set R of vertices that become reachable. (2) Processing the subgraph $<R>$ induced by R. (3) Determining the set X of edges of the form $x \longrightarrow y$, where $x \in R$ and $y \notin R$. (4) Processing the set X of edges.

The set R of vertices that become reachable is obtained easily enough from a simple graph traversal starting from w. Processing the subgraph $<R>$ can be done using a batch algorithm as follows: The dominator tree D of $<R>$, with w as the source vertex, is computed using a batch algorithm. The status of edges in $<R>$ can be computed using a simple traversal of the dominator tree D. The tree D is made a subtree of the original dominator tree by making w a child of u. Priorities can be assigned to vertices in R using a simple batch algorithm—we just need to ensure that the new priorities are all greater than $priority(u)$.

The above steps are sufficient as long as there is no edge from a vertex in R to some previously reachable vertices. If there is such an edge, we identify the set X of edges from vertices in R to vertices outside R. Now we need to process the "insertion" of the edges in X using the special case algorithm outlined previously. We could process these edges one by one, but that is not necessary. We know that

$$\bigcup_{x \longrightarrow y \in X} PossibleAffected(x,y)$$ is an approximation to the set of affected vertices.

This approximation to the affected vertices can be processed as before.

The overall complexity of the algorithm consists of two components: the time spent on updating priorities, which is $O(\gamma \log \gamma)$,[3] and the time spent on the remaining steps, which is $O(\| VISITED \| \log n)$, where $VISITED$ is the approximation to AFFECTED identified by the algorithm.

Identifying Irreducibility
The algorithm discussed so far assumes that the new graph is reducible. In general, it would be useful to determine if the insertion of a new edge can introduce irreducibility in the flowgraph. This test can be done without increasing the asymptotic complexity of the edge-insertion algorithm as follows.

---

[3]The Alpern *et al.* algorithm can take time $O(\gamma^2 \log \gamma)$ to update priorities after the insertion of a number of edges into the dag. The special situation that arises in our algorithm is effectively equivalent to the insertion of a single edge and can be processed in $O(\gamma \log \gamma)$ time.

---

procedure InsertEdge$_{DominatorTree}$$(G, u \rightarrow w)$
declare
      G: a directed graph
      $u \rightarrow w$: an edge to be inserted into G
begin
[1]   Insert edge $u \rightarrow w$ into $E(G)$
[2]   **if** $reachable(u)$ **then**
[3]   **if not** $reachable(w)$ **then**
[4]   Compute the set of newly reachable vertices R and mark them as reachable
[5]   Compute the dominator tree D of <R>, the subgraph induced by R, with w as
               the source vertex
[6]   Make D a subtree of u by performing $link(u,w)$
[7]   Determine the status of all edges in <R>
[8]   Correctly prioritize the dag induced by the forward edges of <R>,
[9]   ensuring that the priorities assigned are greater than $priority(u)$
[10]   Insert all forward edges in <R> into $F(G)$
[11]   Compute X, the set of edges from vertices in R to vertices outside R
[12]   **else**
[13] $X := \{ u \rightarrow w \}$
[14]   **fi**
[15]   Determine the status of all edges in X
[16]   Insert all the forward edges in X into $F(G)$ and update the prioritization of $F(G)$
[17] $S := \underset{x \rightarrow y \,\in\, X}{\cup} PossibleAffected(x,y)$
[18]   **for** every $x \in S$ in increasing order of priority **do**
[19] $y :=$ the least common ancestor of all predecessors in $F(G)$ of x
[20]   **if** $y \ne idom(x)$ **then**
[21] $cut(idom(x),x)$
[22] $link(y,x)$
[23]   **fi**
[24]   **od**
[25] **fi**
end

Figure 9.8. An outline of the algorithm for updating the dominator tree of a reducible flow-graph after the insertion of an edge $v \rightarrow w$ into graph G.

    If the newly inserted edge is a forward edge, then we need to determine if adding this edge to the dag of forward edges of the original flowgraph creates a cycle. We get this information for free, since the Alpern et al. priority updating algorithm (invoked in line [16]) will automatically determine if adding the edges to $F(G)$ will create a cycle in $F(G)$.

    However, this is not sufficient. An inserted edge can be correctly classified as a backward or forward edge using domination information for the original graph, regardless of whether the original and new graph are reducible or not. The status of all other edges in the original graph remain unchanged *if both the original and new graph are reducible*. However, it is possible to insert a forward edge into a reducible flowgraph that changes the status of an original backedge into a forward edge. Such an insertion introduces irreducibility and may not be detected by the test for cycles described above (since the collection of forward edges in the original graph and the newly inserted forward edge may together induce only an acyclic graph).

In order to identify such an edge insertion, we have to check that no change of edge status from "back edge" to "forward edge" occurs. This is not too difficult. At the end of the processing, one just has to check back edges of the form $z \rightarrow x$, where x is in the shaded region of Figure 9.6 to make sure that they are still back edges. (That is, we verify that x is still an ancestor of z in the new dominator tree.) This adds only a constant factor to the time complexity of the algorithm.

9.4.2. Deletion of an Edge

We now consider the problem of updating the dominator tree and the auxiliary information after the deletion of an edge $u \rightarrow w$. Again, we first consider the special case where the edge-deletion does not change the reachability status of any vertex. The vertex w becomes unreachable following the deletion of $u \rightarrow w$ iff $u \rightarrow w$ was the only forward edge coming into w.

A Special Case: No Change in Reachability

We know there is no change in the reachability status of w if there is some other incoming forward edge at w. In this case, we remove the edge $u \rightarrow w$ from $F(G)$ if the edge is a forward edge. Since the previous prioritization of $F(G)$ continues to be a correct prioritization of $F(G)$, no updating of priorities is required. The only non-trivial work is in updating the dominator tree.

We can very easily determine a good approximation to the set of affected vertices after the deletion of the edge $v \rightarrow w$.

Proposition 9.4. Consider the deletion of an edge $u \rightarrow w$ from a flowgraph, where both u and w remain reachable in the new graph too. If a vertex y is affected by the deletion of the edge, then y must be a sibling of w in the original dominator tree—that is, $idom(y) = idom(w)$ in the original dominator tree.

Proof. Assume we obtain graph G_2 from graph G_1 by deleting edge $u \rightarrow w$. The insertion of the deleted edge back into G_2 will restore the dominator tree to its original form. In other words, the set of affected vertices when $u \rightarrow w$ is inserted into G_2 is the same as the set of affected vertices when $u \rightarrow w$ is deleted from G_1. We know from Proposition 9.3 that if a vertex y is affected by the insertion of edge $u \rightarrow w$ into G_2 then it must be a sibling of w in the dominator tree for G_1. $\square$

The General Case

The general-case updating algorithm for processing the deletion of an edge $u \rightarrow w$ is similar to the general-case updating algorithm for processing an edge-insertion. The algorithm performs the following steps: (1) Determine the set R of vertices that have become unreachable, (2) Determine the set F of edges of the form

procedure DeleteEdge$_{DominatorTree}$($G, u \rightarrow w$)
declare
        G: a directed graph
        $u \rightarrow w$: an edge to be deleted from G
begin
[1]   Remove edge $u \rightarrow w$ from $E(G)$
[2]   **if** $u \rightarrow w$ is a forward edge **then**
[3]   Remove edge $u \rightarrow w$ from $F(G)$
[4]   **if** w has no incoming forward edges **then**
[5]   Let R be the descendants of w in the dominator tree
[6]   Mark vertices in R unreachable
[7]   Let X be the set of forward edges from vertices in R to vertices outside R
[8]   Remove X and all forward edges in $<R>$ from $F(G)$
[9]   **else**
[10] $X := \{ u \rightarrow w \}$
[11]   **fi**
[12] $S := \bigcup_{x \rightarrow y \in X} Siblings(y)$
[13]   **for** every $x \in S$ in increasing order of priority **do**
[14] $y :=$ the least common ancestor of all predecessors in $F(G)$ of x
[15]   **if** $y \neq idom(x)$ **then**
[16] $cut(idom(x),x)$
[17] $link(y,x)$
[18]   **fi**
[19]   **od**
[20]   **fi**
end

Figure 9.9. An outline of the algorithm for updating the dominator tree of a reducible flow-graph after the deletion of an edge $v \rightarrow w$ from graph G.

$x \rightarrow y$, where $x \in R$ and $y \notin R$, and (3) Process the set F of edges, by essentially working as though these edges have been deleted from the graph.

    Note that the set R of vertices that have become unreachable is precisely the set of vertices dominated by w, that is, the set of vertices in the subtree T rooted at w in the original dominator tree. The set F of edges is essentially the set of edges $x \rightarrow y$ with x in the subtree T and y not in T—this set of edges is obtained easily using a traversal of the subtree T. The "deletion" of the edges in F is processed as follows: $\bigcup_{x \rightarrow y \in F} Siblings(y)$ is an approximation to the set of affected vertices, which can be processed using previously explained techniques.

    The overall complexity of the edge-deletion procedure is $O(\| VISITED \| \log n)$, where $VISITED$ is the approximation to the set of affected vertices used by the algorithm.

9.4.3. Related Work

The only other incremental algorithm we are aware of for updating the dominator tree is due to Carroll and Ryder.[4] We now briefly compare these two algorithms and argue that the new algorithm is more efficient than the Carroll-Ryder algorithm.

We now briefly compare our incremental algorithm with the incremental algorithm presented by Carroll and Ryder [Car88, Car88a] and argue that the new algorithm is more efficient than the Carroll-Ryder algorithm.

    Our algorithm has a better worst-case time complexity than the Carroll-Ryder algorithm. In the worst case, the algorithm presented in this paper can take $O(m \log n)$ time, where m is the number of edges and n is the number of vertices in the flowgraph. Note that m is $O(n)$ for typical flowgraphs. Carroll and Ryder do not describe the worst-case complexity of their algorithm in terms of n and m. In Carroll's thesis, the complexity of the Carroll-Ryder incremental algorithm is described in terms of "rotations" (see below): their algorithm can perform $\Omega(n^2)$ rotations in the worst case (even with sparse flowgraphs), and the cost of a rotation in the worst-case is $O((f+1) \times g \times d^2)$, where f is the maximum number of children of any vertex in the dominator tree, g is the maximum degree of any vertex in the control-flow graph augmented with certain "representative edges" (see below), and d is the height of the dominator tree. In the worst case, f, g, and d can each be $O(n)$, though f will usually be a small constant for typical flowgraphs. It is not obvious if the worst case for the number of rotations and the parameters f, g, and h can be simultaneously achieved, but the above clearly demonstrates that our algorithm has a better worst-case complexity.

    However, worst-case analysis with the complexity expressed as a function of the (current) input size often does not characterize the complexity of incremental algorithms accurately, and the relative merits of incremental algorithms cannot be established by directly comparing their worst-case complexity measures. For example, the linear-time batch algorithm for constructing the dominator tree has a better complexity than both our incremental algorithm and the Carroll-Ryder incremental algorithm, which have a non-linear worst-case complexity, though the incremental algorithms are arguably better than the batch algorithm. In his thesis, Carroll presents experimental results showing that the Carroll-Ryder algorithm is faster than the almost-linear batch algorithm due to Lengauer and Tarjan [Len79].

    There are two reasons why we believe that our algorithm will perform better in practice than the Carroll-Ryder algorithm. The first is that the Carroll-Ryder algorithm utilizes certain "representative edges" as auxiliary information, and maintaining this auxiliary information can be expensive both in time and space. (For every edge $u \rightarrow v$ in the flowgraph, and for every vertex w that dominates u but not v, a

---

[4]Subsequent to the work described in this book, Sreedhar et al. [Sre95] developed another incremental algorithm for the same problem.

representative edge $w \rightarrow v$ needs to be maintained.) For example, a single edge-deletion (from a sparse flowgraph) that is processed in $O(\log n)$ time by our algorithm can result in the introduction of $\Omega(n^2)$ new representative edges, forcing the Carroll-Ryder algorithm to take $\Omega(n^2)$ time. The second reason is that the Carroll-Ryder algorithm restructures the dominator tree using "local rotation" operations, which move a subtree up or down one level at a time in the dominator tree. (In particular, an upward rotation makes a vertex the child of its original grandparent, while a downward rotation makes a vertex the child of one of its original siblings.) In contrast, our algorithm determines the new immediate dominator of vertices and moves each subtree only once, but such a move can take $O(\log n)$ time because of the use of link and cut operations. Consequently, one can easily create input modifications that are processed in $O(\log n)$ time by our algorithm for which the Carroll-Ryder algorithm performs $\Omega(n)$ rotations, each rotation requiring $\Omega(n)$ time.

One of the differences between our approach and the approach of Carroll and Ryder is that to handle graphs with unreachable vertices they extend the definition of dominator tree to that of a *dominator forest*: they consider a decomposition of the whole graph into a collection of flowgraphs, each with its own source vertex, such that every vertex is reachable from the source of the flowgraph it belongs to; the collection of the dominator trees of these flowgraphs constitute a dominator forest. If the graph has a unique minimal decomposition, then this approach is meaningful. However, a graph need not have a unique minimal decomposition, in general, and the dominator forest is not uniquely defined. Since the advantages of maintaining such a dominator forest are unclear, we have restricted our attention to the problem of maintaining the dominator tree of the reachable vertices.

Finally, we should mention that the algorithm described in Carroll's thesis is capable of handling the simultaneous insertion and deletion of edges from the flowgraph. The algorithm presented in this paper can be adapted to handle the deletion of multiple edges or the insertion of multiple edges. An arbitrary change can be handled by processing all the deleted edges in one step, and then processing all the inserted edges in a second step. Further work is required to explore if the algorithm can be adapted to handle the insertion and deletion of edges simultaneously.

9.4.4. Some Remarks

An interesting direction for future research is suggested by a generalization of the link-cut tree data structure due to Cohen and Tamassia [Coh92, Coh92a]. Consider a tree in which each vertex is associated with some attributes, each of which is defined in terms of attributes of adjacent vertices. Cohen and Tamassia show how to efficiently maintain such an attributed tree dynamically when the attribute equations are linear expressions involving operators that form a semiring. Their algorithm performs link and cut operations in $O(\log n)$ time. Attribute values are not maintained but can be computed on demand in logarithmic time. Carroll and Ryder show that monotonic dataflow analysis problems can be reduced to an attribute evaluation problem over the dominator tree, and use Reps's [Rep82] incremental attribute updating algorithm to perform incremental dataflow analysis. Since the function spaces of distributive dataflow analysis frameworks form a semiring with respect to function composition and meet, it seems worthwhile exploring the possibility of using the Cohen and Tamassia algorithm, in conjunction with our incremental dominator tree algorithm, to perform incremental dataflow analysis.

Chapter 10
Conclusions

> *... the efficiency measure chosen suggests the approach to be taken in tackling an algorithmic problem and guides the development of a solution.*
> —R.E. Tarjan, *1986 Turing Award Lecture*

In 1982, Reps [Rep82] proposed an algorithm for incremental attribute evaluation for noncircular attribute grammars, which he characterized as being "asymptotically optimal" because it ran in time $O(\|\delta\|)$. Subsequently, other optimal algorithms were given for a variety of attribute-grammar subclasses, *e.g.*, absolutely noncircular grammars [Rep84] and ordered attribute grammars [Yeh83, Rep88].

In 1989, Alpern *et al.* [Alp90], addressing a generalization of the attribute updating problem, showed that the dynamic circuit value annotation problem had a lower bound of $\Omega(2^{\|\delta\|})$ under a particular model of computation. They also presented an $O(\|\delta\| \log \|\delta\|)$ algorithm for the problem of updating priorities in a dag, and used it in an unbounded incremental algorithm for the circuit value annotation problem.

In 1991, Ramalingam and Reps outlined "almost optimal", $O(\|\delta\| \log \|\delta\|)$, incremental algorithms for the problem of updating shortest-path information for a graph, a problem that arose in a networking application. (The algorithms presented in Chapter 4 are adaptations and extensions of these algorithms.)

Almost no other incremental graph algorithm had, meanwhile, been characterized as being "optimal" or "almost optimal" in this sense.[1] The complexity of no other algorithm we were aware of had been analyzed in terms of the parameter $\|\delta\|$.

This led us to ask the question "Why?", and to the work described in this book.

Most of the results presented in this book were motivated by the desire to describe the complexity of incremental algorithms in terms of the parameter $\|\delta\|$. This book presents an interesting hierarchy for incremental computation that emerges from this desire to measure complexity in terms of $\|\delta\|$. On the more practical side, the book presents several useful incremental algorithms.

Undeniably, the complexity measure used can guide the development of an algorithm. An implication of this is that the complexity measure used can, sometimes, fail to guide the algorithm designer in the right direction (if it is a wrongly chosen measure). The guiding role played by the complexity measure in algorithm

---

[1] Except for other optimal incremental algorithms for several variants of the attribute updating problem.

design is illustrated by the following examples, drawn from Chapter 7, where we discussed the dynamic circuit value annotation problem:

- It was the desire for a bounded incremental algorithm that suggested the breadth-first iterative evaluate-and-expand strategy for this problem. The complexity measure, worst-case analysis in terms of $\| \delta \|$, however, fails to distinguish between *BF_Expansion* and *RBF_Expansion*, while common sense suggests that *RBF_Expansion* is the better algorithm. The author, in fact, originally chose to discuss *BF_Expansion* because the analysis of the algorithm and the proof of boundedness were easier for that algorithm.[2]

- Analyzing the complexity of the algorithms as a function of the number of vertices they visit suggested the double-and-evaluate improvement outlined in *Balanced_BF_Expansion*.

- The use of worst-case analysis, on the other hand, almost led the author to ignore an improvement, the one outlined in Remark 7.2. This improvement was relegated to the status of a mere remark as it did not improve the asymptotic worst-case complexity in any way. Yet, this was probably the most effective improvement to the basic strategy, at least for the particular problem of updating attributes in a Pascal editor, as evidenced by experimental results.

Though a complexity measure can help even in the process of algorithm design, its more conventional role is to provide us with a good idea about the performance characteristics of algorithms. It should enable us to compare algorithms and determine which is a better algorithm if, in fact, one of the algorithms is significantly better than the other. It should ideally, though not necessarily, enable us to determine if an algorithm is "optimal", one that leaves very little scope for improvement.

Summarizing the discussion in Section 2.3, an analytic complexity measure is likely to better fulfill its role as described above when it presents a reasonable approximation to the time the algorithm takes to process most or all input instances, than when it describes only the time the algorithm takes to process some small fraction of all input instances. An analytic complexity measure can achieve this only if it is expressed as a function of the parameter(s) of the input that really determine the running time of the algorithm.

The difficulty in analyzing incremental algorithms, especially those that require solving a number of subproblems, such as the dynamic dominator tree algorithm presented in Chapter 9, is, perhaps, that the running time depends on a large number of parameters.

There appears to be no silver bullet, no single approach to complexity analysis of incremental algorithms that is universally applicable. Each approach to complexity analysis has the potential to make its own contribution to the development of algo-

---

[2]This is no longer true for the current proof.

rithms for any particular problem. Each has its own limitations and scope in terms of answering the above questions. Each has a domain of applicability, where it serves its role well.

Analysis in terms of $\| \delta \|$ is one such approach.

> *... The subjectively last step comes just before; it is to finish the book itself—to stop writing. That's hard.*
>
> *There is always something left undone, always either something more to say, or a better way to say something, or, at the very least, a disturbing vague sense that the perfect addition or improvement is just around the corner, and the dread that its omission would be everlasting cause for regret.*
>
> ...
>
> *Don't wait and hope for one more result, and don't keep on polishing. Even if you do get that result or do remove that sharp corner, you'll only discover another mirage just ahead.*
>
> —P.R. Halmos, *How to write mathematics*

BIBLIOGRAPHY

Abm88.
Abmann, W., "A short review of high speed compilation," pp. 1-10 in *Compiler compilers and high speed compilation, Proceedings of the 2nd CCHSC Workshop* (Berlin, GDR, October 1988), *Lecture Notes in Computer Science, Vol. 371*, ed. D. Hammer, (1988).

Abr87.
Abramsky, S. and Hankin, C., "An introduction to abstract interpretation," pp. 9-31 in *Abstract Interpretation of Declarative Languages*, ed. S. Abramsky and C. Hankin,Ellis Horwood Limited, Chichester, West Sussex, UK (1987).

Agr83.
Agrawal, R. and Detro, K.D., "An efficient incremental LR parser for grammars with epsilon productions," *Acta Informatica* **19** pp. 369-376 (Sept 1983).

Aho86.
Aho, A.V., Sethi, R., and Ullman, J.D., *Compilers: Principles, Techniques, and Tools*, Addison-Wesley, Reading, MA (1986).

Ahu89.
Ahuja, R.K., Magnanti, T.L., and Orlin, J.B., "Network Flows," pp. 211-369 in *Handbooks in OR & MS, Vol. 1*, ed. G.L. Nemhauser et al., (1989).

Alb90.
Alblas, H., "Concurrent incremental attribute evaluation," in *Attribute Grammars and Their Applications, Lecture Notes in Computer Science, Vol. 461*, ed. P. Deransart and M. Jourdan,Springer-Verlag (1990).

Alb91.
Alblas, H. and Melichar, B., *Proceedings of the international summer school on Attribute Grammars, applications and systems*, (Prague, Czechoslovakia, June 4-13, 1991), *Lecture Notes in Computer Science*, Springer-Verlag, New York, NY (1991).

Alp89.
Alpern, B., Carle, A., Rosen, B., Sweeney, P., and Zadeck, K., "Graph attribution as a specification paradigm," *Proceedings of the ACM SIGSOFT/SIGPLAN Software Engineering Symposium on Practical Software Development Environments*, (Boston, MA, November 28-30, 1988), *ACM SIGPLAN Notices* **24**(2) pp. 121-129 (February 1989).

Alp90.
Alpern, B., Hoover, R., Rosen, B.K., Sweeney, P.F., and Zadeck, F.K., "Incremental evaluation of computational circuits," pp. 32-42 in *Proceedings of the First Annual ACM-SIAM Symposium on Discrete Algorithms*, (San Francisco, CA, Jan. 22-24, 1990), Society for Industrial and Applied Mathematics, Philadelphia, PA (1990).

App91.
Appelbe, B., Smith, K., and Stirewalt, K., "PATCH—a new algorithm for rapid incremental dependence analysis," in *Supercomputing*, ed. E.S. Davidson and F. Hossfield,ACM Press (1991).

Aus90.
Ausiello, G., Italiano, G.F., Spaccamela, A.M., and Nanni, U., "Incremental algorithms for

minimal length paths," pp. 12-21 in *Proceedings of the First Annual ACM-SIAM Sympo-sium on Discrete Algorithms*, (San Francisco, CA, Jan. 22-24, 1990), Society for Industrial and Applied Mathematics, Philadelphia, PA (1990).

Aus91.
Ausiello, G., Italiano, G.F., Spaccamela, A.M., and Nanni, U., "Incremental algorithms for minimal length paths," *Journal of Algorithms*, (12) pp. 615-638 (1991).

Aus92.
Ausiello, G., Italiano, G.F., and Nanni, U., "Optimal traversal of directed hypergraphs," TR-92-073 (September 1992).

Bal88.
Ballance, R.A., Butcher, J., and Graham, S.L., "Grammatical abstraction and incremental syntax analysis in a language-based editor," *Proceedings of the ACM SIGPLAN 88 Conference on Programming Language Design and Implementation*, (Atlanta, GA, June 22-24, 1988), *ACM SIGPLAN Notices* **23**(7) pp. 185-198 (July 1988).

Bee91.
Beetem, J.F. and Beetem, A.F., "Incremental scanning and parsing with Galaxy," *IEEE Transactions on Software Engineering* **17**(7) pp. 641-651 (July 1991).

Ber90.
Berman, A.M., Paull, M.C., and Ryder, B.G., "Proving relative lower bounds for incre-mental algorithms," *Acta Informatica* **27** pp. 665-683 (1990).

Ber92.
Berman, A.M., "Lower and upper bounds for incremental algorithms," Ph.D. dissertation and Tech. Rep. DCS-TR-292, Rutgers University, New Brunswick, NJ (October, 1992).

Bin91.
Binkley, D., "Multi-procedure program integration," Ph.D. dissertation and Tech. Rep. TR-1038, Computer Sciences Department, University of Wisconsin, Madison, WI (August 1991).

Bor79.
Borning, A.H., "ThingLab—A constraint-oriented simulation laboratory," Ph.D. disserta-tion, Comp. Sci. Dept., Stanford University, and Tech. Rep. SSL-79-3, Xerox Palo Alto Research Center, Palo Alto, CA (July 1979).

Bri79.
Bricklin, D. and Frankston, B., *VisiCalc Computer Software Program for the Apple II and II Plus*, Personal Software, Inc., Sunnyvale, CA (1979).

Bro88.
Brooks, K.P., "A two-view document editor with user-definable document structure," Technical Report 33, DEC Systems Research Center, Palo Alto, CA (November 1988).

Bur87.
Burke, M., "An interval-based approach to exhaustive and incremental interprocedural data flow analysis," Res. Rep. RC 12702, IBM T.J. Watson Research Center, Yorktown Heights, NY (April 1987).

Bur87a.
Burke, M. and Ryder, B., "Incremental iterative data flow analysis algorithms," Res. Rep. RC 13170, IBM T.J. Watson Research Center, Yorktown Heights, NY (October 1987).

Bur90a.
Burke, M., "An interval-based approach to exhaustive and incremental interprocedural data-flow analysis," *ACM Trans. Program. Lang. Syst.* **12**(3) pp. 341-395 (July 1990).

Bur93.
Burke, M. and Torczon, L., "Interprocedural optimization: Eliminating unnecessary recompilation," *ACM Trans. Program. Lang. Syst.* **15**(3)(July 1993).

Bur90.
Burke, M.G. and Ryder, B.G., "A critical analysis of incremental iterative dataflow analysis algorithms," *IEEE Transactions on Software Engineering* **16**(7) pp. 723-728 (July 1990).

Car87.
Carpenter, C.W. and Horowitz, M., "Generating incremental VLSI compaction spacing constraints," pp. 291-297 in *Design Automation 24th ACM/IEEE Conference Proceedings*, ed. A. O'Neill and D. Thomas,ACM Press (1987).

Car89.
Carpenter, C.W., *Incremental VLSI Compaction (Computer Aided Design)*, Stanford University, Stanford, CA (1989).

Car71.
Carre, B. A., "An algebra for network routing problems," *J. Inst. Maths. Applics.* **7** pp. 273-294 (1971).

Car88.
Carroll, M. and Ryder, B., "Incremental data flow update via attribute and dominator updates," pp. 274-284 in *Conference Record of the Fifteenth ACM Symposium on Principles of Programming Languages*, (San Diego, CA, January 13-15, 1988), ACM, New York, NY (1988).

Car88a.
Carroll, M.D., "Data flow update via dominator and attribute updates," Ph.D. dissertation, Rutgers University, New Brunswick, NJ (May 1988).

Cha81.
Chamberlin, D.D., King, J.C., Slutz, D.R., Todd, S.J.P., and Wade, B.W., "JANUS: An interactive system for document composition," *Proceedings of the ACM SIGPLAN/SIGOA Symposium on Text Manipulation*, (Portland, OR, June 8-10, 1981), *ACM SIGPLAN Notices* **16**(6) pp. 82-91 (June 1981).

Cha87.
Chamberlin, D.D., "Document convergence in an interactive formatting system," *IBM Systems Journal* **31**(1) pp. 58-72 (January 1987).

Che88a.
Chen, P., Harrison, M.A., and Minakata, I., "Incremental document formatting," pp. 93-100 in *Document Processing Systems, Proceedings of the ACM Conference*, ed. A. Solem,ACM Press (1988).

Che88.
Chen, P. and Harrison, M.A., "Multiple representation document development," *IEEE Computer* **21**(1) pp. 15-31 (January 1988).

Che88b.
Chen, P., "A multiple-representation paradigm for document development," Ph.D. dissertation and Tech. Rep. UCB/CSD 88/436, Dept. of Electrical Engineering and Computer Science, University of Californi a–Berkeley, Berkeley, CA (1988).

Che76.
Cheston, G.A., "Incremental algorithms in graph theory," Ph.D. dissertation and Tech. Rep. 91, Dept. of Computer Science, University of Toronto, Toronto, Canada (March 1976).

Che82.
Cheston, G.A. and Corneil, D.G., "Graph property update algorithms and their application to distance matrices," *INFOR* **20**(3) pp. 178-201 (August 1982).

Coh92a.
Cohen, R.F., "Combine and Conquer," Ph.D. dissertation, Technical Report No. CS-92-45, Department of Computer Science, Brown University, Providence, RI (October 1992).

Coh92.
Cohen, R.F. and Tamassia, R., "Combine and Conquer," Technical Report No. CS-92-19, Department of Computer Science, Brown University, Providence, RI (April 1992).

Coo86.
Cooper, K.D. and Kennedy, K., "The impact of interprocedural analysis and optimization in the Rn programming environment," *ACM Trans. Program. Lang. Syst.* **8**(4) pp. 491-523 (October 1986).

Cor90.
Cormen, T.H., Leiserson, C.E., and Rivest, R.L., *Introduction to Algorithms,* MIT Press, Cambridge, MA (1990).

Cou77.
Cousot, P. and Cousot, R., "Abstract interpretation: A unified lattice model for static analysis of programs by construction or approximation of fixpoints," pp. 238-252 in *Conference Record of the Fourth ACM Symposium on Principles of Programming Languages,* (Los Angeles, CA, January 17-19, 1977), ACM, New York, NY (1977).

Cro85.
Crowe, M., Nicol, C., Hughes, M., and Mackay, D., "On converting a compiler into an incremental compiler," *ACM SIGPLAN Notices* **20**(10) pp. 14-22 (Oct 1985).

Cyt89.
Cytron, R., Ferrante, J., Rosen, B.K., Wegman, M.N., and Zadeck, K., "An efficient method of computing static single assignment form," pp. 25-35 in *Conference Record of the Sixteenth ACM Symposium on Principles of Programming Languages,* (Austin, TX, Jan. 11-13, 1989), ACM, New York, NY (1989).

Cyt91.
    Cytron, R., Ferrante, J., Rosen, B.K., Wegman, M.N., and Zadeck, F.K., "Efficiently com-
    puting static single assignment form and the control dependence graph," *ACM Trans. Pro-
    gram. Lang. Syst.* **13**(4) pp. 451-490 (October 1991).

Dem81.
    Demers, A., Reps, T., and Teitelbaum, T., "Incremental evaluation for attribute grammars
    with application to syntax-directed editors," pp. 105-116 in *Conference Record of the
    Eighth ACM Symposium on Principles of Programming Languages,* (Williamsburg, VA,
    Jan. 26-28, 1981), ACM, New York, NY (1981).

Der90.
    Deransart, P. and Jourdan, M., "," in *Proceedings of the International Workshop on Attri-
    bute Grammars and their App lications,* (Paris, France, September 19-21, 1990), *Lecture
    Notes in Comput er Science,* Vol. 461, Springer-Verlag, New York, NY (1990).

de86.
    de Kleer, J., "An assumption-based TMS," *Artificial Intelligence* **28** pp. 127-162 (1986).

de86a.
    de Kleer, J., "Extending the ATMS," *Artificial Intelligence* **28** pp. 163-196 (1986).

de86b.
    de Kleer, J., "Problem solving with the ATMS," *Artificial Intelligence* **28** pp. 197-224
    (1986).

Die87.
    Dietz, P.F. and Sleator, D.D., "Two algorithms for maintaining order in a list," pp.
    365-372 in *Proceedings of the 19th ACM Symposium on Theory of Computing,* (May
    1987).

Dij59.
    Dijkstra, E.W., "A note on two problems in connexion with graphs," *Numerische
    Mathematik* **1** pp. 269-271 (1959).

Dio78.
    Dionne, R., "Etude et extension d'un algorithme de Murchland," *INFOR* **16**(2) pp.
    132-146 (June 1978).

Di89.
    Di Battista, G. and Tamassia, R., "Incremental planarity testing," pp. 436-441 in *Proceed-
    ings of the Thirtieth IEEE Symposium on Foundations of Computer Science,* IEEE Com-
    puter Society, Washington, DC (1989).

Dji95.
    Djidjev, H.N., Pantziou, G.E., and Zaroliagis, C.D., "On-line and Dynamic Algorithms for
    Shortest Path Problems," pp. 193-204 in *Proceedings of Twelfth Annual Symposium on
    Theoretical Aspects of Computer Science,* (Munich, Germany, March, 1995), *Lecture
    Notes in Computer Science,* Vol. 900, ed. E.W. Mayr and C. Puech,Springer-Verlag, New
    York, NY (1995).

Doy79a.
    Doyle, J., "A glimpse of truth maintenance," in *Artificial Intelligence: An M.I.T. Perspec-
    tive,* ed. P.H. Winston and R.H. Brown,The M.I.T. Press, Cambridge, MA (1979).

Doy79.
Doyle, J., "A truth maintenance system," *Artificial Intelligence* **12** pp. 231-272 (1979).

Dri88.
Driscoll, J.R., Gabow, H.N., Shrairman, R., and Tarjan, R.E., "Relaxed heaps: An alternative to Fibonacci heaps with applications to parallel computation," *Communications of the ACM* **31**(11) pp. 1343-1354 (1988).

Edm72.
Edmonds, J. and Karp, R.M., "Theoretical improvements in algorithmic efficiency for network flow problems," *J. ACM* **19** pp. 248-264 (1972).

Epp92a.
Eppstein, D., Galil, Z., Italiano, G.F., and Nissenzweig, A., "Sparsification – A technique for speeding up dynamic graph algorithms," in *Proceedings of the Thirty-third IEEE Symposium on Foundations of Computer Science,* (Pittsburgh, PA, Oct. 25-27, 1992), IEEE Computer Society, Washington, DC (1992).

Epp92.
Eppstein, D., Italiano, G.F., Tamassia, R., Tarjan, R.E., and Westbrook, J., "Maintenance of a minimum spanning forest in a dynamic planar graph," *Journal of Algorithms* **13** pp. 33-54 (1992).

Epp93.
Eppstein, D., Galil, Z., and Italiano, G.F., "Improved Sparsification," Tech. Report 93-20, Department of Inf. and Comp. Science, University of California–Irvine, Irvine, CA (1993).

Eve81.
Even, S. and Shiloach, Y., "An on-line edge-deletion problem," *J. ACM* **28**(1) pp. 1-4 (January 1981).

Eve85.
Even, S. and Gazit, H., "Updating distances in dynamic graphs," pp. 271-388 in *IX Symposium on Operations Research,* (Osnabrueck, W. Ger., Aug. 27-29, 1984), *Methods of Operations Research,* Vol. 49, ed. P. Brucker and R. Pauly,Verlag Anton Hain (1985).

Fen90.
Feng, A., Kikuno, T., and Torii, K., "Incremental attribute evaluation for multiple subtree replacements in structure-oriented environments," in *Attribute Grammars and Their Applications, Lecture Notes in Computer Science, Vol. 461,* ed. P. Deransart and M. Jourdan,Springer-Verlag (1990).

Fil87.
File, G., "Classical and incremental attribute evaluation by means of recursive procedures," *Theoretical Computer Science* **53**(1) pp. 25-65 (1987).

Fis88.
Fischer, C.N. and LeBlanc, R.J., *Crafting a Compiler,* Benjamin/Cummings Publishing Company, Inc., Menlo Park, CA (1988).

Flo62.
Floyd, R.W., "Algorithm 97: shortest path," *Commun. of the ACM* **5** p. 345 (1962).

For85.
Ford, R. and Sawamiphakdi, D., "A greedy concurrent approach to incremental code generation," pp. 165-178 in *Conference Record of the Twelfth ACM Symposium on Principles
of Programming Languages,* (New Orleans, LA, Jan. 14-16, 1985), ACM, New York, NY
(1985).

Fre85.
Frederickson, G.N., "Data structures for on-line updating of minimum spanning trees, with
applications," *SIAM J. Comput.* **14**(4) pp. 781-798 (November 1985).

Fre87.
Fredman, M.L. and Tarjan, R.E., "Fibonacci heaps and their uses in improved network
optimization algorithms," *J. ACM* **34**(3) pp. 596-615 (1987).

Fre90a.
Fredman, M.L. and Willard, D.E., "Trans-dichotomous algorithms for minimum spanning
trees and shortest paths," pp. 719-725 in *Proceedings of the 31st Annual Symposium on
Foundations of Computer Science Volume II* (St. Louis, Missouri, October 1990), IEEE
Computer Society, Washington, DC (1990).

Fre90.
Freeman-Benson, B.N, Maloney, J., and Borning, A., "An incremental constraint solver,"
Commun. of the ACM **33**(1) pp. 54-63 (January 1990).

Fri94.
Frigioni, D., Marchetti-Spaccamela, A., and Nanni, U., "Incremental algorithms for the
single-source shortest path problem," pp. 113-124 in *Proceedings of the fourteenth
FST&TCS Conference, Lecture Notes in Computer Science Vol. 880,* ed. P.S. Thiagarajan,
(1994).

Fri83.
Fritzson, P., "Symbolic debugging through incremental compilation in an integrated
environment," *Journal of Systems and Software* **3**(4) pp. 285-294 (Dec 1983).

Fri84.
Fritzson, P., "Preliminary experience from the DICE system, a distributed incremental
compiling environment," *Proceedings of the ACM SIGSOFT/SIGPLAN Software
Engineering Symposium on Practical Software Development Environments,* (Pittsburgh,
PA, Apr. 23-25, 1984), *ACM SIGPLAN Notices* **19**(5) pp. 113-123 (May 1984).

Fri84a.
Fritzson, P., "Towards a distributed programming environment based on incremental com-
pilation," Linköping Studies in Science and Technology Dissertation No. 109, Dept. of
Comp. and Inf. Sci., Linköping University, Linköping, Sweden (1984).

Fuj88.
Fujita, H. and Furukawa, K., "A self-applicable partial evaluator and its use in incremental
compilation," *New Generation Computing* **6**(2-3) pp. 91-118 (1988).

Gaf90.
    Gafter, N.M., *Parallel incremental compilation,* University of Rochester, Rochester, NY (1990).

Ghe79.
    Ghezzi, C. and Mandrioli, D., "Incremental parsing," *ACM Trans. Program. Lang. Syst.* **1**(1) pp. 58-70 (July 1979).

Ghe80.
    Ghezzi, C. and Mandrioli, D., "Augmenting parsers to support incrementality," *Journal of the ACM* **27**(3) pp. 564-579 (October 1980).

Gho83.
    Ghodssi, V., "Incremental analysis of programs," Ph.D. dissertation, Dept. of Computer Science, University of Central Florida, Orlando, FL (1983).

Gon84a.
    Gondran, M. and Minoux, M., *Graphs and Algorithms,* John Wiley and Sons, New York (1984).

Gon84.
    Gondran, M. and Minoux, M., "Linear algebra in dioids: a survey of recent results," pp. 147-163 in *Algebraic and Combinatorial Methods in Operations Research, Annals of Discrete Mathematics 19,* ed. R.E. Burkard, R.A. Cunninghame-Green, and U. Zimmermann,North-Holland, Amsterdam (1984).

Got78.
    Goto, S. and Sangiovanni-Vincentelli, A., "A new shortest path updating algorithm," *Networks* **8**(4) pp. 341-372 (1978).

Gra76.
    Graham, S. and Wegman, M., "A fast and usually linear algorithm for global data flow analysis," *J. ACM* **23** pp. 172-202 (1976).

Hal70.
    Halder, A.K., "The method of competing links," *Transportation Science* **4** pp. 36-51 (1970).

Ham88.
    Hammer, D., *Compiler compilers and high speed compilation, Lecture Notes in Computer Science, Vol. 371.* 1988.
Har85.
    Harel, D., "A linear time algorithm for finding dominators in flow graphs and related problems," pp. 185-194 in *Proceedings of the Symposium on Theory Of Computing,* (1985).

Har89.
    Harrison, M.A. and Munson, E.V., "On integrated bibliography processing," *Electronic Publishing* **2**(4) pp. 193-210 (December 1989).

Har91.
    Harrison, M.A. and Munson, E.V., "Numbering document components," *Electronic Publishing* **4**(1)(January 1991).

Hee90.
Heeman, F.C., "Incremental parsing of expressions," *Journal of Systems and Software* **13** pp. 55-69 (Sept 1990).

Hen90.
Hentenryck, P.V., "Incremental constraint satisfaction in logic programming," pp. 189-202 in *Logic Programming*, ed. D.H.D. Warren and P. Szeredi,MIT Press (1990).

Hoo86a.
Hoover, R., "Dynamically bypassing copy rule chains in attribute grammars," pp. 14-25 in *Conference Record of the Thirteenth ACM Symposium on Principles of Programming Languages,* (St. Petersburg, FL, Jan. 13-15, 1986), ACM, New York, NY (1986).

Hoo86.
Hoover, R. and Teitelbaum, T., "Efficient incremental evaluation of aggregate values in attribute grammars," *Proceedings of the SIGPLAN 86 Symposium on Compiler Construction,* (Palo Alto, CA, June 25-27, 1986), *ACM SIGPLAN Notices* **21**(7) pp. 39-50 (July 1986).

Hoo87.
Hoover, R., "Incremental graph evaluation," Ph.D. dissertation and Tech. Rep. 87-836, Dept. of Computer Science, Cornell University, Ithaca, NY (May 1987).

Hoo92.
Hoover, R., "Alphonse: Incremental Computation as a Programming Abstraction," *Proceedings of the ACM SIGPLAN 92 Conference on Programming Language Design and Implementation,* (San Francisco, CA, June 17-19, 1992), *ACM SIGPLAN Notices* **27**(7) pp. 261-272 (July 1992).

Hsi76.
Hsieh, W., Kershenbaum, A., and Golden, B., "Constrained routing in large sparse networks," pp. 38.14-38.18 in *Proceedings of IEEE International Conference on Communications,* , Philadelphia, PA (1976).

Hud91.
Hudson, S.E., "Incremental attribute evaluation: a flexible algorithm for lazy update," *ACM Trans. Program. Lang. Syst.* **13**(3) pp. 315-341 (July 1991).

Ita86.
Italiano, G.F., "Amortized efficiency of a path retrieval data structure," *Theoretical Computer Science* **48** pp. 273-281 (1986).

Ita88.
Italiano, G.F., "Finding paths and deleting edges in directed acyclic graphs," *Information Processing Letters* **28** pp. 5-11 (1988).

Jab88.
Jaber, A.M., *Development of an incremental parser of LALR(1) languages,* Lehigh University, Bethlehem, PA (1988).

Jai90.
Jain, A.N. and Waibel, A.H., "Incremental parsing by modular recurrent connectionist networks," pp. 364-371 in *Advances in neural information processing systems 2,* ed. D.S.

Touretzky,Morgan-Kaufman,
Palo Alto, CA (1990).

Jal82.
Jalili, F. and Gallier, J., "Building friendly parsers," pp. 196-206 in *Conference Record of the Ninth ACM Symposium on Principles of Programming Languages,* (Albuquerque, NM, Jan. 25-27, 1982), ACM, New York, NY (1982).

Joh77.
Johnson, D.B., "Efficient algorithms for shortest paths in sparse networks," *JACM* **24**(1) pp. 1-13 (1977).

Joh82.
Johnson, G.F. and Fischer, C.N., "Non-syntactic attribute flow in language based editors," pp. 185-195 in *Conference Record of the Ninth annual ACM Symposium on Principles Of Programming Languages,* (Albuquerque, NM, January 25-27, 1982), ACM, New York, NY (1982).

Joh83.
Johnson, G.F., "An approach to incremental semantics," Ph.D. dissertation, Computer Sciences Department, University of Wisconsin, Madison, WI (December 1983).

Joh85.
Johnson, G.F. and Fischer, C.N., "A meta-language and system for nonlocal incremental attribute evaluation in language-based editors," pp. 141-151 in *Conference Record of the Twelfth ACM Symposium on Principles of Programming Languages,* (New Orleans, LA, Jan. 14-16, 1985), ACM, New York, NY (1985).

Jon86.
Jones, L. and Simon, J., "Hierarchical VLSI design systems based on attribute grammars," pp. 58-69 in *Conference Record of the Thirteenth ACM Symposium on Principles of Programming Languages,* (St. Petersburg, FL, Jan. 13-15, 1986), ACM, New York, NY (1986).

Jon90.
Jones, L.G., "Efficient evaluation of circular attribute grammars," *ACM Trans. Program. Lang. Syst.* **12**(3) pp. 429-462 (July 1990).

Kai85.
Kaiser, G.E. and Kant, E., "Incremental parsing without a parser," *Journal of Systems and Software* **5**(2) pp. 121-144 (May 1985).

Kai89.
Kaiser, G.E., "Incremental dynamic semantics for language-based programming environments," *ACM Trans. Program. Lang. Syst.* **11**(2) pp. 169-193 (April 1989).

Kam76.
Kam, J.B. and Ullman, J.D., "Global data flow analysis and iterative algorithms," *J. ACM* **23** pp. 158-171 (1976).

Kam77.
Kam, J.B. and Ullman, J.D., "Monotone data flow analysis frameworks," *Acta Informatica* **7** pp. 305-317 (1977).

Kap86.
   Kaplan, S. and Kaiser, G., "Incremental attribute evaluation in distributed language-based editors," pp. 121-130 in *Proceedings of the Fifth ACM Symposium on Principles of Distributed Computing*, (1986).

Kil73.
   Kildall, G., "A unified approach to global program optimization," pp. 194-206 in *Conference Record of the First ACM Symposium on Principles of Programming Lan guages*, ACM, New York, NY (1973).

Knu73.
   Knuth, D.E., *The Art of Computer Programming, Vol. 1: Fundamental Algorithms*, Addison-Wesley, Reading, MA (1968, Second Edition: 1973).

Knu77.
   Knuth, D.E., "A generalization of Dijkstra's algorithm," *Information Processing Letters* **6**(1) pp. 1-5 (1977).

Knu81.
   Knuth, D.E. and Plass, M.F., "Breaking paragraphs into lines," *Software—Practice and Experience* **11** pp. 1119-1184 (1981).

Kon84.
   Konopasek, M. and Jayaraman, S., *The TK!Solver Book*, Osborne/McGraw-Hill, Berkeley, CA (1984).

Ku89.
   Ku, C.S., *Incremental compilation of rules in indefinite deductive databases*, Northwestern University, Evanston, IL (1989).

Len79.
   Lengauer, T. and Tarjan, R.E., "A fast algorithm for finding dominators in a flowgraph," *ACM Transactions on Programming Languages and Systems* **1**(1) pp. 121-141 (July 1979).

Len91.
   Lengauer, T. and Theune, D., "Unstructured path problems and the making of semirings," pp. 189-200 in *Proceedings of the second Workshop on Algorithms and Data Structures* (Ottawa, Ontario), (1991).

Lin90.
   Lin, C.-C. and Chang, R.-C., "On the dynamic shortest path problem," *Journal of Information Processing* **13**(4)(1990).

Lin89.
   Linton, M.A. and Quong, R.W., "A macroscopic profile of program compilation and linking," *IEEE Transactions on Software Engineering* **15**(4) pp. 427-436 (April 1989).

Lou67.
   Loubal, P., "A network evaluation procedure," *Highway Research Record* **205** pp. 96-109 (1967).

Mah84.
   Mahr, B., "Iteration and summability in semirings," pp. 229-256 in *Algebraic and Combinatorial Methods in Operations Research, Annals of Discrete Mathematics 19,* ed. R.E. Burkard, R.A. Cunninghame-Green, and U. Zimmermann,North-Holland, Amsterdam (1984).

Man88.
   Mannila, H. and Ukkonen, E., "Time parameter and arbitrary deunions in the set union problem," pp. 34-42 in *Proceedings of the First Scandinavian Workshop on Algorithm Theory (SWAT 88), Lecture Notes in Computer Science,* Vol. 318, Springer-Verlag, New York, NY (1988).

Mar92.
   Marchetti-Spaccamela, A., Nanni, U., and Rohnert, H., *On-line Graph Algorithms for Incremental Compilation,* TR-92-056, International Computer Science Institute, Berkeley, CA (1992).

Mar89.
   Marlowe, T.J., "Data flow analysis and incremental iteration," Ph.D. dissertation and Tech. Rep. DCS-TR-255, Rutgers University, New Brunswick, NJ (October 1989).

Mar90.
   Marlowe, T.J. and Ryder, B.G., "An efficient hybrid algorithm for incremental data flow analysis," pp. 184-196 in *Conference Record of the Seventeenth ACM Symposium on Principles of Programming Languages,* (San Francisco, CA, Jan. 17-19, 1990), ACM, New York, NY (1990).

Mar90a.
   Marlowe, T.J. and Ryder, B.G., "Properties of data flow framework: A unified model," *Acta Informatica* **28** pp. 121-163 (1990).

McA90.
   McAllester, D., "Truth maintenance," pp. 92-104 in *Proceedings of the Eighth National Conference on Artificial Intelligence,* (Boston, MA, July 29 – August 3, 1990), AAAI Press/The M.I.T. Press, Cambridge, MA (1990).

Mur92.
   Murata, M. and Hayashi, K., "Formatter hierarchy for structured documents," pp. 77-94 in *Proceedings of Electronic Publishing, 1992,* ed. C. Vanoirbeek and G. Coray,Cambridge University Press, Cambridge, Great Britain (1992).

Mur67.
   Murchland, J.D., "The effect of increasing or decreasing the length of a single arc on all shortest distances in a graph," Tech. Rep. LBS-TNT-26, London Business School, Transport Network Theory Unit, London, UK (1967).

Mur.
   Murchland, J.D., "A fixed matrix method for all shortest distances in a directed graph and for the inverse problem," Doctoral dissertation, Universität Karlsruhe, Karlsruhe, W. Germany ().

Och83.
   Ochranova, R., "Finding dominators," pp. 328-334 in *Proceedings of the Foundations of Computation Theory,* (1983).

Ous84a.
Ousterhout, J.K., "Corner stitching: A data-structuring technique for VLSI layout tools," *IEEE Transactions on Computer-Aided Design* **CAD-3**(1) pp. 87-100 (January 1984).

Ous84.
Ousterhout, J.K., Hamachi, G.T., Mayo, R.N., Scott, W.S., and Taylor, G.S., "Magic: A VLSI layout system," pp. 152-159 in *Proceedings of the Twenty-First Design Automation Conference*, IEEE Computer Society, Washington, DC (1984).

Pap74.
Pape, U., "Netzwerk-veraenderungen und korrektur kuerzester weglaengen von einer wurzelmenge zu allen anderen knoten," *Computing* **12** pp. 357-362 (1974).

Par83.
Pardo, R.K. and Landau, R., "Process and apparatus for converting a source program into an object program," U.S. Patent No. 4,398,249, United States Patent Office, Washington, DC (August 9, 1983).

Par88.
Parigot, D., "Transformations, évaluation incrémentale et optimizations des grammai es attribuées: Le système FNC-2," These de Doctorat, L'Université Paris XI, Centre D'Orsay, Orsay, France (1988).

Pec90.
Peckham, S.B., "Incremental attribute evaluation and multiple subtree replacements," Ph.D. dissertation, Dept. of Computer Science, Cornell University, Ithaca, NY (1990).

Per84.
Perlis, D., "Bibliography of literature on non-monotonic reasoning," (Source unknown), (1984).

Pol85.
Pollock, L.L. and Soffa, M.L., "Incremental compilation of optimized code," pp. 152-164 in *Conference Record of the Twelfth ACM Symposium on Principles of Programming Languages,* (New Orleans, LA, Jan. 14-16, 1985), ACM, New York, NY (1985).

Pol86.
Pollock, L.L., *An approach to incremental compilation of optimized code,* University of Pittsburgh, Pittsburgh, PA (1986).

Pol92.
Pollock, L.L. and Soffa, M.L., "Incremental global reoptimization of programs," *ACM Trans. Program. Lang. Syst.* **14**(2) pp. 173-200 (April 1992).

Pou88.
Poutre, J.A. La and Leeuwen, J. van, "Maintenance of transitive closures and transitive reductions of graphs," pp. 106-120 in *Graph-Theoretic Concepts in Computer Science: Proceedings of the 14th International Workshop*(1988), *Lecture Notes in Computer Science,* Vol. 314, (1988).

Pug88.
  Pugh, W.W., "Incremental computation and the incremental evaluation of functional pro-
  grams," Ph.D. dissertation and Tech. Rep. 88-936, Dept. of Computer Science, Cornell
  University, Ithaca, NY (August 1988).

Quo91.
  Quong, R.W. and Linton, M.A., "Linking programs incrementally," *ACM Trans. Program. Lang. Syst.* **13**(1) pp. 1-20 (Jan 1991).

Ram91.
  Ramalingam, G. and Reps, T., "On the computational complexity of incremental algo-
  rithms," TR-1033, Computer Sciences Department, University of Wisconsin, Madison,
  WI (August 1991).

Ram93.
  Ramalingam, G. and Reps, T., "A categorized bibliography on incremental computation,"
  pp. 502-510 in *In the Conference Record of the Twentieth ACM Symposium on Principles
  of Programming Languages* (Charleston, SC, January 11-13, 1993)., ACM, New York,
  NY (1993). (Tutorial paper.)

Ram94.
  Ramalingam, G. and Reps, T., "An incremental algorithm for maintaining the dominator
  tree of a reducible flowgraph," pp. 287-296 in *Conference Record of the Twenty-First
  ACM Symposium on Principles of Programming Languages,* (Portland, OR, January 17-
  21, 1994)., ACM, New York, NY (1994).

Ram96.
  Ramalingam, G. and Reps, T., "On the computational complexity of dynamic graph problems," To appear in *Theoretical Computer Science A* **162**(July 1996).

Ram.
  Ramalingam, G. and Reps, T., "An incremental algorithm for a generalization of the
  shortest-path problem," To appear in *Journal of Algorithms*, ().

Rei87.
  Reif, J.H., "A topological approach to dynamic graph connectivity," *Information Processing Letters* **25**(1) pp. 65-70 (1987).

Rei84.
  Reiss, S., "An approach to incremental compilation," *Proceedings of the SIGPLAN 84
  Symposium on Compiler Construction,* (Montreal, Can., June 20-22, 1984), *ACM SIG-
  PLAN Notices* **19**(6) pp. 144-156 (June 1984).

Rep82.
  Reps, T., "Optimal-time incremental semantic analysis for syntax-directed editors," pp.
  169-176 in *Conference Record of the Ninth ACM Symposium on Principles of Programming Languages,* (Albuquerque, NM, January 25-27, 1982), ACM, New York, NY
  (1982).

Rep83.
  Reps, T., Teitelbaum, T., and Demers, A., "Incremental context-dependent analysis for
  language-based editors," *ACM Trans. Program. Lang. Syst.* **5**(3) pp. 449-477 (July 1983).

Rep84.
Reps, T., *Generating Language-Based Environments,* The M.I.T. Press, Cambridge, MA (1984).

Rep86.
Reps, T., Marceau, C., and Teitelbaum, T., "Remote attribute updating for language-based editors," pp. 1-13 in *Conference Record of the Thirteenth ACM Symposium on Principles of Programming Languages,* (St. Petersburg, FL, Jan. 13-15, 1986), ACM, New York, NY (1986).

Rep88a.
Reps, T., "Incremental evaluation for attribute grammars with unrestricted movement between tree modifications," *Acta Informatica,* pp. 155-178 (1988).

Rep88.
Reps, T. and Teitelbaum, T., *The Synthesizer Generator: A System for Constructing Language-Based Editors,* Springer-Verlag, New York, NY (1988).

Rep88b.
Reps, T. and Teitelbaum, T., *The Synthesizer Generator Reference Manual: Third Edition,* Springer-Verlag, New York, NY (1988).

Rod68.
Rodionov, V., "The parametric problem of shortest distances," *U.S.S.R. Computational Math. and Math. Phys.* **8**(5) pp. 336-343 (1968).

Roh85.
Rohnert, H., "A dynamization of the all pairs least cost path problem," pp. 279-286 in *Proceedings of STACS 85: Second Annual Symposium on Theoretical Aspects of Computer Science,* (Saarbruecken, W. Ger., Jan. 3-5, 1985), *Lecture Notes in Computer Science,* Vol. 182, ed. K. Mehlhorn,Springer-Verlag, New York, NY (1985).

Ros81.
Rosen, B.K., "Linear cost is sometimes quadratic," pp. 117-124 in *Conference Record of the Eighth ACM Symposium on Principles of Programming Languages,* (Williamsburg, VA, January 26-28, 1981), ACM, New York, NY (1981).

Ros90.
Rosene, C.M., "Incremental dependence analysis," Ph.D. dissertation and Tech Rep. CRPC-TR90044, Center for Research on Parallel Computation, Rice University, Houston, TX (March 1990).

Rot90.
Rote, G., "Path problems in graphs," pp. 155-189 in *Computational Graph Theory (Computing Supplementum 7),* ed. G. Tinhofer et al.,Springer-Verlag, New York, NY (1990).

Ryd82.
Ryder, B., "Incremental data flow analysis based on a unified model of elimination algorithms," Ph.D. dissertation and Tech. Rep. DCS-TR-117, Rutgers University, New Brunswick, NJ (September 1982).

Ryd88.
Ryder, B.G. and Paull, M.C., "Incremental data flow analysis algorithms," *ACM Trans. Program. Lang. Syst.* **10**(1) pp. 1-50 (January 1988).

Ryd90.
　　Ryder, B.G., Landi, W., and Pande, H.D., "Profiling an incremental data flow analysis algorithm," *IEEE Transactions on Software Engineering* **SE-16**(2)(February 1990).

Sai93.
　　Sairam, S., Vitter, J.S., and Tamassia, R., *A complexity-theoretic approach to incremental computation.* 1993.

Sch88.
　　Schwanke, R.W. and Kaiser, G.E., "Technical Correspondence: Smarter recompilation," *ACM Trans. Program. Lang. Syst.* **10**(4) pp. 627-632 (October 1988).

Sch84.
　　Schwartz, M., Delisle, N., and Begwani, V., "Incremental compilation in Magpie," *Proceedings of the SIGPLAN 84 Symposium on Compiler Construction,* (Montreal, Can., June 20-22, 1984), *ACM SIGPLAN Notices* **19**(6) pp. 122-131 (June 1984).

Sco84.
　　Scott, W.S. and Ousterhout, J.K., "Plowing: Interactive stretching and compaction in Magic," pp. 166-172 in *Proceedings of the Twenty-First Design Automation Conference,* IEEE Computer Society, Washington, DC (1984).

Sed83.
　　Sedgewick, R., *Algorithms,* Addison-Wesley, Reading, MA (1983).

Sha88.
　　Shanahan, M., "Incrementality and logic programming," pp. 21-34 in *Reason maintenance systems and their applications*, ed. B. Smith and G. Kelleher,Halstead Press, New York, NY (1988).

Sha93.
　　Shao, Z. and Appel, A.W., "Smartest Recompilation," pp. 439-450 in *Conference Record of the Twentieth ACM Symposium on Principles of Programming Languages,* (Charleston, SC, Jan. 10-13, 1993), ACM, New York, NY (1993).

Sha81.
　　Sharir, M. and Pnueli, A., "Two approaches to interprocedural data flow analysis," pp. 189-233 in *Program Flow Analysis: Theory and Applications*, ed. S.S. Muchnick and N.D. Jones,Prentice-Hall, Englewood Cliffs, NJ (1981).

Shm84.
　　Shmueli, O., Tsur, S., and Zfira, H., "Rule support in Prolog," (Source unknown), (1984).

Shm90.
　　Shmueli, O. and Tsur, S., "Incremental re-evaluation of LDL queries," pp. 99-111 in *Logic Programming*, ed. D.H.D. Warren and P. Szeredi,MIT PRess, Cambridge, MA (1990).

Sle83.
　　Sleator, D.D and Tarjan, R.E., "A data structure for dynamic trees," *Journal of Computer and System Sciences* **26** pp. 362-391 (1983).

Smi88.
   Smith, B. and Kelleher, G., *Reason maintenance systems and their applications*, Halstead Press, New York, NY (1988).

Spi75.
   Spira, P.M. and Pan, A., "On finding and updating spanning trees and shortest paths," *SIAM J. Computing* **4**(3) pp. 375-380 (September 1975).

Sre95.
   Sreedhar, V.C., Gao, G.R., and Lee, Y., "Incremental computation of dominator trees," pp. 1-12 in *In the Proceedings of the ACM SIGPLAN Workshop on Intermediate Representation, (*Also in *SIGPLAN Notices, Vol. 30, Number 4)*, (January 1995).

Sto77.
   Stoy, J.E., *Denotational Semantics: The Scott-Strachey Approach to Programming Language Theory*, The M.I.T. Press, Cambridge, MA (1977).

Sud92.
   Sudarshan, S., "Optimizing bottom-up query evaluation for deductive databases," Ph.D. dissertation and Tech. Rep. TR-1125, Computer Sciences Department, University of Wisconsin, Madison, WI (November 1992).

Tan85.
   Tan, Z. and Lemone, K.A., "A research environment for incremental data flow analysis," pp. 356-362 in *Proceedings of the 1985 ACM Computer Science Conference*, (1985).

Tar81a.
   Tarjan, R.E., "Fast algorithms for solving path problems," *J. ACM* **28** pp. 594-614 (1981).

Tar81.
   Tarjan, R.E., "A unified approach to path problems," *J. ACM* **28** pp. 576-593 (1981).

Tar83.
   Tarjan, R.E., *Data Structures and Network Algorithms*, Society for Industrial and Applied Mathematics, Philadelphia, PA (1983).

Tay88.
   Taylor, D., "An incremental compilation package in C," *Computer Language* **5**(2) pp. 59-70 (Feb 1988).

Tay84.
   Taylor, G.S. and Ousterhout, J.K., "Magic's incremental design-rule checker," pp. 160-165 in *Proceedings of the Twenty-First Design Automation Conference*, IEEE Computer Society, Washington, DC (1984).

Tei90.
   Teitelbaum, T. and Chapman, R., "Higher-order attribute grammars and editing environments," *Proceedings of the ACM SIGPLAN 90 Conference on Programming Language Design and Implementation*, (White Plains, NY, June 20-22, 1990), *ACM SIGPLAN Notices* **25**(6) pp. 197-208 (June 1990).

Tic86.
   Tichy, W.F., "Smart recompilation," *ACM Trans. Program. Lang. Syst.* **8**(3) pp. 637-654 (July 1986).

Tic88.

Tichy, W.F., "Technical Correspondence: Tichy's response to R.W. Schwanke and G.E. Kaiser's "Smarter recompilation"," *ACM Trans. Program. Lang. Syst.* **10**(4) pp. 633-634 (October 1988).

Van88.

Vander Zanden, B. T., "Incremental constraint satisfaction and its application to graphical interfaces," Ph.D. dissertation and Tech. Rep. TR 88-941, Dept. of Computer Science, Cornell University, Ithaca, NY (October 1988).

Vor90.

Vorthmann, S.A., "Coordinated incremental attribute evaluation on a DR-threaded tree," in *Attribute Grammars and Their Applications, Lecture Notes in Computer Science, Vol. 461,* ed. P. Deransart and M. Jourdan,Springer-Verlag (1990).

Vor90a.

Vorthmann, S.A., *Syntax-directed editor support for incremental consistency maintenance,* Georgia Institute of Technology, Atlanta, GA (1990).

Wal88.

Walz, J.A. and Johnson, G.F., "Incremental evaluation for a general class of circular attribute grammars," *Proceedings of the ACM SIGPLAN 88 Conference on Programming Language Design and Implementation,* (Atlanta, GA, June 22-24, 1988), *ACM SIGPLAN Notices* **23**(7) pp. 209-221 (July 1988).

Wan.

Wang, F.-J., *A high-level graph approach to incremental data flow analysis,* Northwestern University, Evanston, IL ().

Weg80.

Wegman, M., "Parsing for structural editors," pp. 320-327 in *Proceedings of the Twenty-First IEEE Symposium on Foundations of Computer Science* (Syracuse, NY, October 1980), IEEE Computer Society, Washington, DC (1980).

Wir92.

Wiren, M., "Studies in incremental natural-language analysis," Linkoping Studies in Science and Technology. Dissertation No. 292, Linkoping University, Sweden (1992).

Wir93.

Wiren, M., "Bounded Incremental Parsing," in *Proceedings of the Twente Workshop on Language Technology (TWLT 6) (Natural Language Parsing: Methods and Formalisms),* (December 1993).

Wol91.

Wolfson, O., Dewan, H.M., Salvatore, S.J., and Yemini, Y., "Incremental evaluation of rules and its relationship to parallelism," *ACM SIGMOD Record* **20**(2) pp. 78-87 (June 1991).

Yan90.

Yannakakis, M., "Graph-theoretic methods in database theory," pp. 230-242 in *Proceedings of the Ninth ACM Symposium on Principles of Database Systems,* (Nashville, Tennessee, April 2-4, 1990), ACM, New York, NY (1990).

Yap83.
  Yap, C.K., "A hybrid algorithm for the shortest path between two nodes in the presence of few negative arcs," *Information Processing Letters* **16** pp. 181-182 (May 1983).

Yeh83.
  Yeh, D., "On incremental evaluation of ordered attributed grammars," *BIT* **23** pp. 308-320 (1983).

Yeh88.
  Yeh, D. and Kastens, U., "Improvements of an incremental evaluation algorithm for ordered attribute grammars," *ACM SIGPLAN Notices* **23**(12) pp. 45-50 (Dec 1988).

Yel.
  Yellin, D.M., "Speeding up dynamic transitive closure for bounded degree graphs," To appear in *Acta Informatica*, ().

Zad83.
  Zadeck, F.K., "Incremental data flow analysis in a structured program editor," Ph.D. dissertation, Mathematical Sciences Dept., Rice University, Houston, TX (October 1983).

Zad84.
  Zadeck, F.K., "Incremental data flow analysis in a structured program editor," *Proceedings of the SIGPLAN 84 Symposium on Compiler Construction,* (Montreal, Can., June 20-22, 1984), *ACM SIGPLAN Notices* **19**(6) pp. 132-143 (June 1984).

Zar90.
  Zaring, A., "Parallel evaluation in attribute grammar based systems," Ph.D. dissertation and Tech. Rep. 90-1149, Dept. of Computer Science, Cornell University, Ithaca, NY (August 1990).

Zim81.
  Zimmermann, U., *Linear and Combinatorial Optimization in Ordered Algebraic Structures, Annals of Discrete Mathematics 10,* North-Holland, Amsterdam (1981).

Springer-Verlag
and the Environment

We at Springer-Verlag firmly believe that an international science publisher has a special obligation to the environment, and our corporate policies consistently reflect this conviction.

We also expect our business partners – paper mills, printers, packaging manufacturers, etc. – to commit themselves to using environmentally friendly materials and production processes.

The paper in this book is made from low- or no-chlorine pulp and is acid free, in conformance with international standards for paper permanency.

Lecture Notes in Computer Science

For information about Vols. 1–1019

please contact your bookseller or Springer-Verlag

Vol. 1020: I.D. Watson (Ed.), Progress in Case-Based Reasoning. Proceedings, 1995. VIII, 209 pages. 1995. (Subseries LNAI).

Vol. 1021: M.P. Papazoglou (Ed.), OOER '95: Object-Oriented and Entity-Relationship Modeling. Proceedings, 1995. XVII, 451 pages. 1995.

Vol. 1022: P.H. Hartel, R. Plasmeijer (Eds.), Functional Programming Languages in Education. Proceedings, 1995. X, 309 pages. 1995.

Vol. 1023: K. Kanchanasut, J.-J. Lévy (Eds.), Algorithms, Concurrency and Knowlwdge. Proceedings, 1995. X, 410 pages. 1995.

Vol. 1024: R.T. Chin, H.H.S. Ip, A.C. Naiman, T.-C. Pong (Eds.), Image Analysis Applications and Computer Graphics. Proceedings, 1995. XVI, 533 pages. 1995.

Vol. 1025: C. Boyd (Ed.), Cryptography and Coding. Proceedings, 1995. IX, 291 pages. 1995.

Vol. 1026: P.S. Thiagarajan (Ed.), Foundations of Software Technology and Theoretical Computer Science. Proceedings, 1995. XII, 515 pages. 1995.

Vol. 1027: F.J. Brandenburg (Ed.), Graph Drawing. Proceedings, 1995. XII, 526 pages. 1996.

Vol. 1028: N.R. Adam, Y. Yesha (Eds.), Electronic Commerce. X, 155 pages. 1996.

Vol. 1029: E. Dawson, J. Golić (Eds.), Cryptography: Policy and Algorithms. Proceedings, 1995. XI, 327 pages. 1996.

Vol. 1030: F. Pichler, R. Moreno-Díaz, R. Albrecht (Eds.), Computer Aided Systems Theory - EUROCAST '95. Proceedings, 1995. XII, 539 pages. 1996.

Vol.1031: M. Toussaint (Ed.), Ada in Europe. Proceedings, 1995. XI, 455 pages. 1996.

Vol. 1032: P. Godefroid, Partial-Order Methods for the Verification of Concurrent Systems. IV, 143 pages. 1996.

Vol. 1033: C.-H. Huang, P. Sadayappan, U. Banerjee, D. Gelernter, A. Nicolau, D. Padua (Eds.), Languages and Compilers for Parallel Computing. Proceedings, 1995. XIII, 597 pages. 1996.

Vol. 1034: G. Kuper, M. Wallace (Eds.), Constraint Databases and Applications. Proceedings, 1995. VII, 185 pages. 1996.

Vol. 1035: S.Z. Li, D.P. Mital, E.K. Teoh, H. Wang (Eds.), Recent Developments in Computer Vision. Proceedings, 1995. XI, 604 pages. 1996.

Vol. 1036: G. Adorni, M. Zock (Eds.), Trends in Natural Language Generation - An Artificial Intelligence Perspective. Proceedings, 1993. IX, 382 pages. 1996. (Subseries LNAI).

Vol. 1037: M. Wooldridge, J.P. Müller, M. Tambe (Eds.), Intelligent Agents II. Proceedings, 1995. XVI, 437 pages. 1996. (Subseries LNAI).

Vol. 1038: W. Van de Velde, J.W. Perram (Eds.), Agents Breaking Away. Proceedings, 1996. XIV, 232 pages. 1996. (Subseries LNAI).

Vol. 1039: D. Gollmann (Ed.), Fast Software Encryption. Proceedings, 1996. X, 219 pages. 1996.

Vol. 1040: S. Wermter, E. Riloff, G. Scheler (Eds.), Connectionist, Statistical, and Symbolic Approaches to Learning for Natural Language Processing. IX, 468 pages. 1996. (Subseries LNAI).

Vol. 1041: J. Dongarra, K. Madsen, J. Waśniewski (Eds.), Applied Parallel Computing. Proceedings, 1995. XII, 562 pages. 1996.

Vol. 1042: G. Weiß, S. Sen (Eds.), Adaption and Learning in Multi-Agent Systems. Proceedings, 1995. X, 238 pages. 1996. (Subseries LNAI).

Vol. 1043: F. Moller, G. Birtwistle (Eds.), Logics for Concurrency. XI, 266 pages. 1996.

Vol. 1044: B. Plattner (Ed.), Broadband Communications. Proceedings, 1996. XIV, 359 pages. 1996.

Vol. 1045: B. Butscher, E. Moeller, H. Pusch (Eds.), Interactive Distributed Multimedia Systems and Services. Proceedings, 1996. XI, 333 pages. 1996.

Vol. 1046: C. Puech, R. Reischuk (Eds.), STACS 96. Proceedings, 1996. XII, 690 pages. 1996.

Vol. 1047: E. Hajnicz, Time Structures. IX, 244 pages. 1996. (Subseries LNAI).

Vol. 1048: M. Proietti (Ed.), Logic Program Syynthesis and Transformation. Proceedings, 1995. X, 267 pages. 1996.

Vol. 1049: K. Futatsugi, S. Matsuoka (Eds.), Object Technologies for Advanced Software. Proceedings, 1996. X, 309 pages. 1996.

Vol. 1050: R. Dyckhoff, H. Herre, P. Schroeder-Heister (Eds.), Extensions of Logic Programming. Proceedings, 1996. VII, 318 pages. 1996. (Subseries LNAI).

Vol. 1051: M.-C. Gaudel, J. Woodcock (Eds.), FME'96: Industrial Benefit and Advances in Formal Methods. Proceedings, 1996. XII, 704 pages. 1996.

Vol. 1052: D. Hutchison, H. Christiansen, G. Coulson, A. Danthine (Eds.), Teleservices and Multimedia Communications. Proceedings, 1995. XII, 277 pages. 1996.

Vol. 1053: P. Graf, Term Indexing. XVI, 284 pages. 1996. (Subseries LNAI).

Vol. 1054: A. Ferreira, P. Pardalos (Eds.), Solving Combinatorial Optimization Problems in Parallel. VII, 274 pages. 1996.

Vol. 1055: T. Margaria, B. Steffen (Eds.), Tools and Algorithms for the Construction and Analysis of Systems. Proceedings, 1996. XI, 435 pages. 1996.

Vol. 1056: A. Haddadi, Communication and Cooperation in Agent Systems. XIII, 148 pages. 1996. (Subseries LNAI).

Vol. 1057: P. Apers, M. Bouzeghoub, G. Gardarin (Eds.), Advances in Database Technology — EDBT '96. Proceedings, 1996. XII, 636 pages. 1996.

Vol. 1058: H. R. Nielson (Ed.), Programming Languages and Systems – ESOP '96. Proceedings, 1996. X, 405 pages. 1996.

Vol. 1059: H. Kirchner (Ed.), Trees in Algebra and Programming – CAAP '96. Proceedings, 1996. VIII, 331 pages. 1996.

Vol. 1060: T. Gyimóthy (Ed.), Compiler Construction. Proceedings, 1996. X, 355 pages. 1996.

Vol. 1061: P. Ciancarini, C. Hankin (Eds.), Coordination Languages and Models. Proceedings, 1996. XI, 443 pages. 1996.

Vol. 1062: E. Sanchez, M. Tomassini (Eds.), Towards Evolvable Hardware. IX, 265 pages. 1996.

Vol. 1063: J.-M. Alliot, E. Lutton, E. Ronald, M. Schoenauer, D. Snyers (Eds.), Artificial Evolution. Proceedings, 1995. XIII, 396 pages. 1996.

Vol. 1064: B. Buxton, R. Cipolla (Eds.), Computer Vision – ECCV '96. Volume I. Proceedings, 1996. XXI, 725 pages. 1996.

Vol. 1065: B. Buxton, R. Cipolla (Eds.), Computer Vision – ECCV '96. Volume II. Proceedings, 1996. XXI, 723 pages. 1996.

Vol. 1066: R. Alur, T.A. Henzinger, E.D. Sontag (Eds.), Hybrid Systems III. IX, 618 pages. 1996.

Vol. 1067: H. Liddell, A. Colbrook, B. Hertzberger, P. Sloot (Eds.), High-Performance Computing and Networking. Proceedings, 1996. XXV, 1040 pages. 1996.

Vol. 1068: T. Ito, R.H. Halstead, Jr., C. Queinnec (Eds.), Parallel Symbolic Languages and Systems. Proceedings, 1995. X, 363 pages. 1996.

Vol. 1069: J.W. Perram, J.-P. Müller (Eds.), Distributed Software Agents and Applications. Proceedings, 1994. VIII, 219 pages. 1996. (Subseries LNAI).

Vol. 1070: U. Maurer (Ed.), Advances in Cryptology – EUROCRYPT '96. Proceedings, 1996. XII, 417 pages. 1996.

Vol. 1071: P. Miglioli, U. Moscato, D. Mundici, M. Ornaghi (Eds.), Theorem Proving with Analytic Tableaux and Related Methods. Proceedings, 1996. X, 330 pages. 1996. (Subseries LNAI).

Vol. 1072: R. Kasturi, K. Tombre (Eds.), Graphics Recognition. Proceedings, 1995. X, 308 pages. 1996.

Vol. 1073: J. Cuny, H. Ehrig, G. Engels, G. Rozenberg (Eds.), Graph Grammars and Their Application to Computer Science. Proceedings, 1994. X, 565 pages. 1996.

Vol. 1074: G. Dowek, J. Heering, K. Meinke, B. Möller (Eds.), Higher-Order Algebra, Logic, and Term Rewriting. Proceedings, 1995. VII, 287 pages. 1996.

Vol. 1075: D. Hirschberg, G. Myers (Eds.), Combinatorial Pattern Matching. Proceedings, 1996. VIII, 392 pages. 1996.

Vol. 1076: N. Shadbolt, K. O'Hara, G. Schreiber (Eds.), Advances in Knowledge Acquisition. Proceedings, 1996. XII, 371 pages. 1996. (Subseries LNAI).

Vol. 1077: P. Brusilovsky, P. Kommers, N. Streitz (Eds.), Mulimedia, Hypermedia, and Virtual Reality. Proceedings, 1994. IX, 311 pages. 1996.

Vol. 1078: D.A. Lamb (Ed.), Studies of Software Design. Proceedings, 1993. VI, 188 pages. 1996.

Vol. 1079: Z.W. Raś, M. Michalewicz (Eds.), Foundations of Intelligent Systems. Proceedings, 1996. XI, 664 pages. 1996. (Subseries LNAI).

Vol. 1080: P. Constantopoulos, J. Mylopoulos, Y. Vassiliou (Eds.), Advanced Information Systems Engineering. Proceedings, 1996. XI, 582 pages. 1996.

Vol. 1081: G. McCalla (Ed.), Advances in Artificial Intelligence. Proceedings, 1996. XII, 459 pages. 1996. (Subseries LNAI).

Vol. 1082: N.R. Adam, B.K. Bhargava, M. Halem, Y. Yesha (Eds.), Digital Libraries. Proceedings, 1995. Approx. 310 pages. 1996.

Vol. 1083: K. Sparck Jones, J.R. Galliers, Evaluating Natural Language Processing Systems. XV, 228 pages. 1996. (Subseries LNAI).

Vol. 1084: W.H. Cunningham, S.T. McCormick, M. Queyranne (Eds.), Integer Programming and Combinatorial Optimization. Proceedings, 1996. X, 505 pages. 1996.

Vol. 1085: D.M. Gabbay, H.J. Ohlbach (Eds.), Practical Reasoning. Proceedings, 1996. XV, 721 pages. 1996. (Subseries LNAI).

Vol. 1086: C. Frasson, G. Gauthier, A. Lesgold (Eds.), Intelligent Tutoring Systems. Proceedings, 1996. XVII, 688 pages. 1996.

Vol. 1087: C. Zhang, D. Lukose (Eds.), Distributed Artificial Intelliegence. Proceedings, 1995. VIII, 232 pages. 1996. (Subseries LNAI).

Vol. 1088: A. Strohmeier (Ed.), Reliable Software Technologies – Ada-Europe '96. Proceedings, 1996. XI, 513 pages. 1996.

Vol. 1089: G. Ramalingam, Bounded Incremental Computation. XI, 190 pages. 1996.

Vol. 1090: J.-Y. Cai, C.K. Wong (Eds.), Computing and Combinatorics. Proceedings, 1996. X, 421 pages. 1996.

Vol. 1091: J. Billington, W. Reisig (Eds.), Application and Theory of Petri Nets 1996. Proceedings, 1996. VIII, 549 pages. 1996.

Vol. 1092: H. Kleine Büning (Ed.), Computer Science Logic. Proceedings, 1995. VIII, 487 pages. 1996.